T0213880

SpringerBriefs in Applied Sciences and Technology

SpringerBriefs in Continuum Mechanics

Series Editors

Holm Altenbach, Magdeburg, Germany
Andreas Öchsner, Johor Bahru, Malaysia

For further volumes:
http://www.springer.com/series/10528

Per Olsson

Transport Phenomena in Newtonian Fluids - A Concise Primer

 Springer

Per Olsson
Göteborg
Sweden

ISSN 2191-530X   ISSN 2191-5318 (electronic)
ISBN 978-3-319-01308-4   ISBN 978-3-319-01309-1 (eBook)
DOI 10.1007/978-3-319-01309-1
Springer Cham Heidelberg New York Dordrecht London

Library of Congress Control Number: 2013944552

Printed on acid-free paper

Springer is part of Springer Science+Business Media (www.springer.com)

The original version of the book was revised.
For detailed information please see erratum.
The erratum to the book is available at
https://doi.org/10.1007/978-3-319-01309-1_7

Contents

Introduction

This book describes transport phenomena in Newtonian fluids such as momentum transport, energy transport and mass transport. The book contains detailed derivations of the transport equations for these transport phenomena. The book also contains analytical solutions to the transport equations in some simple geometries.

Chapter 1 is a description of the basic mathematics used in the book. The chapter is not intended to be a textbook of mathematics, but contains only such information which is necessary for the reader to be able to read and understand the book's other content.

Chapter 2, which deals with momentum transport, contains a derivation of the Navier-Stokes-Duhem equation describing flow in a Newtonian fluid. Chapter 2 also contains the derivations of the Bernoulli equation, the pressure equation and the wave equation for sound waves. Further, the chapter contains analytical solutions to the flow equation in some simple geometries. The chapter also describes the boundary layer, turbulent flow and flow separation.

Chapter 3, which deals with energy transport, contains a derivation of the heat transport equation describing heat transport in a flowing Newtonian fluid. Heat transport in a flowing fluid is caused by thermal conduction and convection. The chapter also contains a definition of the heat transfer coefficient and analytical solutions for the heat transfer coefficient in some simple geometries. Chapter 2 contains the solutions to the Navier-Stokes equation in these geometries.

Chapter 4, which deals with mass transport, contains a derivation of the mass transport equation describing mass transport in a flowing Newtonian fluid. Mass transport in a flowing fluid is caused by diffusion and convection. The chapter also contains a definition of the mass transfer coefficient and analytical solutions for the mass transfer coefficient in some simple geometries. Chapter 2 contains the solutions to the Navier-Stokes equation in these geometries.

Chapter 1
Elementary Mathematics

1.1 Introduction

This chapter is intended for readers who are not familiar with the vector and tensor notation appearing in the book. The transport equations become much more compact if they are written with vector or tensor notation. This is especially true when the flow equation is written with tensor notation. The chapter is not intended to be a textbook of mathematics, but contains only such information which is necessary for the reader to be able to read and understand the book's other content. There is something improper to speak about vector and tensor notation. It is more proper to speak about *symbolic* and *indicial* notation but in this book *symbolic notation* will be called *vector notation* and *indicial notation* will be called *tensor notation*. Section. 1.3 contains only very basic information about tensors. The most important in Sect. 1.3 is the *Einstein summation convention* and the way to write partial derivatives with respect to the space coordinates with tensor notation. All quantities which are written with tensor notation in this book are Cartesian tensors. Equations written in other coordinate systems are not written with tensor notation. The chapter also contains descriptions of line integrals, surface integrals, volume integrals and some mathematical theorems such as the Stokes theorem and the Gauss theorem.

1.2 Vector Notation

An example of a vector is the the space vector $\mathbf{x}$ in a Cartesian coordinate system. Vectors are denoted by bold straight style in this book. The space vector $\mathbf{x}$ can be written

$$\mathbf{x} = \begin{bmatrix} x_1 \\ x_2 \\ x_3 \end{bmatrix} = \begin{bmatrix} x \\ y \\ z \end{bmatrix} \tag{1.1}$$

An erratum to this chapter is available at https://doi.org/10.1007/978-3-319-01309-1_5.

P. Olsson, *Transport Phenomena in Newtonian Fluids - A Concise Primer*,
SpringerBriefs in Continuum Mechanics,
DOI 10.1007/978-3-319-01309-1_1, © The Author(s) 2014

Fig. 1.1 The vectors **a** and **b**

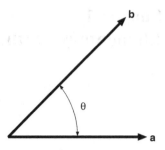

where x, y and z are space coordinates in a Cartesian coordinate system.

1.2.1 Scalar Product

The scalar product between the vectors **a** and **b** is denoted $\mathbf{a} \cdot \mathbf{b}$ and is defined (see Fig. 1.1)

$$\mathbf{a} \cdot \mathbf{b} = \mathbf{b} \cdot \mathbf{a} = \begin{bmatrix} a_1 & a_2 & a_3 \end{bmatrix} \begin{bmatrix} b_1 \\ b_2 \\ b_3 \end{bmatrix}$$

$$= a_1 b_1 + a_2 b_2 + a_3 b_3 = |\mathbf{a}| \, |\mathbf{b}| \cos \theta \qquad (1.2)$$

1.2.2 Cross Product

The cross product between the vectors **a** and **b** is denoted $\mathbf{a} \times \mathbf{b}$ and is defined (see Fig. 1.2)

Fig. 1.2 The vectors **a**, **b** and $\mathbf{a} \times \mathbf{b}$

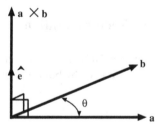

$$\mathbf{a} \times \mathbf{b} = -\mathbf{b} \times \mathbf{a} = \begin{vmatrix} \widehat{\mathbf{x}}_1 & \widehat{\mathbf{x}}_2 & \widehat{\mathbf{x}}_3 \\ a_1 & a_2 & a_3 \\ b_1 & b_2 & b_3 \end{vmatrix}$$

$$= \widehat{\mathbf{x}}_1 \, (a_2 \, b_3 \, - \, a_3 \, b_2) + \widehat{\mathbf{x}}_2 \, (a_3 \, b_1 \, - \, a_1 \, b_3) + \widehat{\mathbf{x}}_3 \, (a_1 \, b_2 \, - \, a_2 \, b_1)$$

$$= \widehat{\mathbf{e}} \, |\mathbf{a}| \, |\mathbf{b}| \sin \theta \tag{1.3}$$

where $\widehat{\mathbf{x}}_1$, $\widehat{\mathbf{x}}_2$ and $\widehat{\mathbf{x}}_3$ are the unit vectors in a Cartesian coordinate system.

1.2.3 The ∇ Operator

The vector operator ∇ (nabla) occurs frequently in various transport equations. ∇ is defined

$$\nabla = \begin{bmatrix} \dfrac{\partial}{\partial x_1} \\ \dfrac{\partial}{\partial x_2} \\ \dfrac{\partial}{\partial x_3} \end{bmatrix} \tag{1.4}$$

1.2.4 Gradient Vector

The gradient ∇f of a scalar function f which is a function of the the space coordinates $\mathbf{x}$ is defined

$$\nabla f = \begin{bmatrix} \dfrac{\partial f}{\partial x_1} \\ \dfrac{\partial f}{\partial x_2} \\ \dfrac{\partial f}{\partial x_3} \end{bmatrix} \tag{1.5}$$

1.2.5 The Laplace Operator

The Laplace operator ∇^2 occurs frequently in many partial differential equations. The Laplace operator is defined

$$\nabla^2 = \nabla \cdot \nabla = \left[\frac{\partial}{\partial x_1} \ \frac{\partial}{\partial x_2} \ \frac{\partial}{\partial x_3} \right] \begin{bmatrix} \frac{\partial}{\partial x_1} \\ \frac{\partial}{\partial x_2} \\ \frac{\partial}{\partial x_3} \end{bmatrix}$$

$$= \frac{\partial^2}{\partial x_1^2} + \frac{\partial^2}{\partial x_2^2} + \frac{\partial^2}{\partial x_3^2} \tag{1.6}$$

1.2.6 Divergence

The divergence of a vector valued function **F** which depends on the space coordinates **x** is denoted $\nabla \cdot \mathbf{F}$ and is defined

$$\nabla \cdot \mathbf{F} = \left[\frac{\partial}{\partial x_1} \ \frac{\partial}{\partial x_2} \ \frac{\partial}{\partial x_3} \right] \begin{bmatrix} F_1 \\ F_2 \\ F_2 \end{bmatrix}$$

$$= \frac{\partial F_1}{\partial x_1} + \frac{\partial F_2}{\partial x_2} + \frac{\partial F_3}{\partial x_3} \tag{1.7}$$

1.2.7 Curl

The curl of a vector valued function **F** which depends on the space coordinates **x** is denoted $\nabla \times \mathbf{F}$ and is defined

Fig. 1.3 The vectors **F**, **x** and $d\mathbf{x}$ at the line C

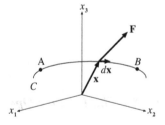

Fig. 1.4 The vector **F** at the surface S bounded by the line C

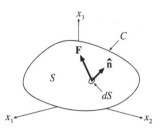

$$\nabla \times \mathbf{F} = \begin{vmatrix} \widehat{\mathbf{x}}_1 & \widehat{\mathbf{x}}_2 & \widehat{\mathbf{x}}_3 \\ \dfrac{\partial}{\partial x_1} & \dfrac{\partial}{\partial x_2} & \dfrac{\partial}{\partial x_3} \\ F_1 & F_2 & F_3 \end{vmatrix}$$

$$= \widehat{\mathbf{x}}_1 \left(\frac{\partial F_3}{\partial x_2} - \frac{\partial F_2}{\partial x_3} \right) + \widehat{\mathbf{x}}_2 \left(\frac{\partial F_1}{\partial x_3} - \frac{\partial F_3}{\partial x_0 1} \right) + \widehat{\mathbf{x}}_3 \left(\frac{\partial F_2}{\partial x_1} - \frac{\partial F_1}{\partial x_2} \right) \tag{1.8}$$

where $\widehat{\mathbf{x}}_1$, $\widehat{\mathbf{x}}_2$ and $\widehat{\mathbf{x}}_3$ are the unit vectors in a Cartesian coordinate system.

1.2.8 Line Integrals

The line integral of a vector valued function **F** along the line C between point A and B in Fig. 1.3 is defined

$$dI_C = \mathbf{F} \cdot d\mathbf{x} \tag{1.9}$$

$$I_C = \int_C \mathbf{F} \cdot d\mathbf{x} = \int_{\mathbf{x}_A}^{\mathbf{x}_B} \mathbf{F} \cdot d\mathbf{x} \tag{1.10}$$

where $d\mathbf{x}$ is an infinitesimal displacement along the line C. I_C is the sum of all infinitesimal scalar products $\mathbf{F} \cdot d\mathbf{x}$ between A and B along the line C.

1.2.9 Surface Integrals

The surface integral of a vector valued function **F** over the surface S which is limited by the line C in Fig. 1.4 is defined

Fig. 1.5 The volume V
enclosed by the surface S

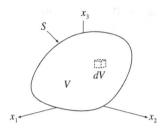

$$dI_S = \mathbf{F} \cdot \hat{\mathbf{n}}\, dS \ = \ \mathbf{F} \cdot d\mathbf{S} \tag{1.11}$$

$$I_S = \int_S \mathbf{F} \cdot d\mathbf{S} \tag{1.12}$$

where dS is an infinitesimal surface element on the surface S. $\hat{\mathbf{n}}$ is the unit normal on the surface S. I_S is the sum of all infinitesimal scalar products $\mathbf{F} \cdot d\mathbf{S}$ on the surface S.

1.2.10   Volume Integrals

The volume integral I_V of a scalar function f in volume V which is enclosed by the surface S in Fig. 1.5 is defined

$$dI_V = f\, dx_1\, dx_2\, dx_3 \ = \ f\, dV \tag{1.13}$$

$$I_V = \int_V f\, dV \tag{1.14}$$

where dV is an infinitesimal volume element. I_V is the sum of all infinitesimal scalars $f\, dV$ in the volume V.

1.2.11   The Stokes Theorem

Assume that the surface S is enclosed by the line C in Fig. 1.4. Then, the line integral of $\mathbf{F}$ along the line C can be written

$$\oint_C \mathbf{F} \cdot d\mathbf{x} = \int_S \nabla \times \mathbf{F} \cdot d\mathbf{S} \tag{1.15}$$

1.2.12   The Gauss Theorem

Assume that the volume V is enclosed by the surface S in Fig. 1.5. Then, the surface integral of $\mathbf{F}$ over the surface S can be written

$$\int_S \mathbf{F} \cdot d\mathbf{S} = \int_V \nabla \cdot \mathbf{F} \, dV \tag{1.16}$$

Equation (1.16) assumes that there are no singularities in volume V.

1.3   Tensor Notation

The word tensor comes from the Latin word *tendo* which means stretch. An example of a tensor is the stress tensor. Tensor notation means that the stress tensor can be written

$$\sigma_{ij} = \begin{bmatrix} \sigma_{11} & \sigma_{12} & \sigma_{13} \\ \sigma_{21} & \sigma_{22} & \sigma_{23} \\ \sigma_{31} & \sigma_{32} & \sigma_{33} \end{bmatrix} \tag{1.17}$$

With vector notation, the stress tensor is written as Σ. See also the introduction to this chapter. Tensor notation means that array variables are written with indices. For example, the space vector $\mathbf{x}$ is written x_i with tensor notation.

$$x_i = \begin{bmatrix} x_1 \\ x_2 \\ x_3 \end{bmatrix} \tag{1.18}$$

An advantage with tensor notation is that there is no problem to write array variables with higher dimension than two. For example, it is difficult to write the array variable a_{ijk} with vector notation. There are also other advantages with tensor notation.

1.3.1   The Einstein Summation Convention

A characteristic rule for tensor notation is that sums are written without summation signs. Summation occurs implicitly for repeated indices. The following examples illustrate this.

$$a_{ii} = \sum_{i=1}^{3} a_{ii} = a_{11} + a_{22} + a_{33} \tag{1.19}$$

$$a_{ij} b_j = \sum_{j=1}^{3} a_{ij} b_j = a_{i1} b_1 + a_{i2} b_2 + a_{i3} b_3 \tag{1.20}$$

This rule is called the *Einstein summation convention*. The rule is valid for all Cartesian tensors in this book.

1.3.2 Derivative Notation

The partial derivative of a scalar function f with respect to the space coordinate x_i is denoted

$$\frac{\partial f}{\partial x_i} = f_{,i} \tag{1.21}$$

The derivative is denoted by a comma followed by the index of the space coordinate. Note that this notation is only used for partial derivatives with respect to the space coordinates x_i. Other derivatives are written in the common way.

1.3.3 The Kronecker Delta

The symbol δ_{ij} appears in equations written with tensor notation. δ_{ij} is called the *Kronecker delta* and is defined

$$\delta_{ij} = \begin{cases} 1 & \text{if } i = j \\ 0 & \text{if } i \neq j \end{cases} \tag{1.22}$$

1.3.4 The Levi-Civita Symbol

The symbol ϵ_{ijk} appears in equations written with tensor notation. ϵ_{ijk} is called the *Levi-Civita symbol* and is defined

$$\epsilon_{ijk} = \begin{cases} 1 & \text{if the values of } i, j, k \text{ are an even permutation of } 1, 2, 3 \\ & \text{(that is, they appear as in the sequence } 1, 2, 3, 1, 2). \\ -1 & \text{if the values of } i, j, k \text{ are an odd permutation of } 1, 2, 3 \\ & \text{(that is, they appear as in the sequence } 3, 2, 1, 3, 2). \\ 0 & \text{if at least two of the values of } i, j, k \text{ are equal.} \end{cases}$$

$$\tag{1.23}$$

1.3.5 Scalar Product

With tensor notation, the scalar product $\mathbf{a} \cdot \mathbf{b}$ between the vectors a_i and b_i can be written

$$\mathbf{a} \cdot \mathbf{b} = a_i \, b_i \;=\; a_1 \, b_1 + a_2 \, b_2 + a_3 \, b_3 \tag{1.24}$$

This is a consequence of the Einstein summation convention.

1.3.6 Cross Product

With tensor notation, the cross product $\mathbf{a} \times \mathbf{b}$ between the vectors a_i and b_i can be written

$$\mathbf{a} \times \mathbf{b} = \epsilon_{ijk} \, a_j \, b_k \tag{1.25}$$

where ϵ_{ijk} is the Levi-Civita symbol. See Eq. (1.23).

1.3.7 Gradient Vector

With tensor notation, the gradient vector ∇f of a scalar function f can be written

$$\nabla f = f_{,i} \tag{1.26}$$

1.3.8 Divergence

With tensor notation, the divergence $\nabla \cdot \mathbf{F}$ of a vector valued function F_i can be written

$$\nabla \cdot \mathbf{F} = F_{i,i} \;=\; \frac{\partial F_1}{\partial x_1} + \frac{\partial F_2}{\partial x_2} + \frac{\partial F_3}{\partial x_3} \tag{1.27}$$

1.3.9 Curl

With tensor notation, the curl $\nabla \times \mathbf{F}$ of a vector valued function F_i can be written

$$\nabla \times \mathbf{F} = \epsilon_{ijk} \, F_{k,j} \tag{1.28}$$

where ϵ_{ijk} is the Levi-Civita symbol. See Eq. (1.23).

1.3.10   Line Inegrals

With tensor notation, the line integral of a vector valued function F_i can be written

$$\int_C \mathbf{F} \cdot d\mathbf{x} = \int_C F_i \, dx_i \qquad (1.29)$$

1.3.11   Surface Integrals

With tensor notation, the surface integral of a vector valued function F_i can be written

$$\int_S \mathbf{F} \cdot d\mathbf{S} = \int_S F_i \, n_i \, dS \qquad (1.30)$$

where n_i is the unit normal on the surface S.

1.3.12   The Stokes Theorem

With tensor notation, the Stokes theorem can be written

$$\oint_C F_i \, dx_i = \int_S \epsilon_{ijk} \, F_{k,\, j} \, n_i \, dS \qquad (1.31)$$

where n_i is the unit normal on the surface S and ϵ_{ijk} is the Levi-Civita symbol. See Eq. (1.23).

1.3.13   The Gauss Theorem

With tensor notation, the Gauss theorem can be written

$$\int_S F_i \, n_i \, dS = \int_V F_{i,\, i} \, dV \qquad (1.32)$$

where n_i is the unit normal on the surface S.

Chapter 2
Momentum Transport

2.1 Introduction

This chapter describes the flow equation which is valid for a Newtonian fluid. This equation is called the *Navier-Stokes-Duhem equation*. The chapter also describes some special cases of the Navier-Stokes-Duhem equation such as the *Navier-Stokes equation* for incompressible flow, the *Bernoulli equation*, the *pressure equation* and the *wave equation*. These equations are often written with tensor notation in the chapter but sometimes even with vector notation. Note that the Einstein summation convention is valid for tensors. This means that summation occurs implicitly for repeated indices. The derivative of a function f with respect to the space coordinate x_i is denoted $f_{,i}$ with tensor notation. The chapter also includes some examples of analytical solutions to the flow equation in some simple geometries. In these cases, the acceleration forces or the friction forces in the flow are neglected or equal to zero. It can be very complicated to solve the flow equation analytically if the acceleration forces or the friction forces are not zero even for very simple geometries. An example of this is radial flow in a slit. In this case, the flow equation can be solved analytically for creeping flow or frictionless flow. Generally, the flow equation is included in a system of equations which also includes equations for energy and mass transport. In many special cases, one can solve the flow equation separately without taking into account the other equations.

2.2 The Navier-Stokes-Duhem Equation

Newton's equation means that the momentum balance over a fixed closed surface S which encloses the volume V in a Newtonian fluid can be written

---

The original version of this chapter was revised: For detailed information please see Erratum. An erratum to this chapter can be found at https://doi.org/10.1007/978-3-319-01309-1_7.

P. Olsson, *Transport Phenomena in Newtonian Fluids - A Concise Primer*, SpringerBriefs in Continuum Mechanics, DOI 10.1007/978-3-319-01309-1_2, © The Author(s) 2014

$$\frac{\partial}{\partial t} \int_V \rho \, v_i \, dV \; - \; \int_V \rho \, g_i \, dV \; = \; - \int_S \rho \, v_i \, v_j \, n_j \, dS \; + \; \int_S \sigma_{ji} \, n_j \, dS \qquad (2.1)$$

$$\sigma_{ji} = \sigma_{ij} = - \, p \, \delta_{ij} + \tau_{ij} \qquad (2.2)$$

$$\tau_{ij} = \lambda \, \delta_{ij} \, v_{k,k} + \mu \, (v_{i,j} + v_{j,i}) \qquad v_{i,j} + v_{j,i} = 2 \, D_{ij} \qquad (2.3)$$

Equation (2.1) is the equation of motion expressed as an integral equation. The first term on the left side is the time derivative of the momentum in the volume V. The second integral on the left side is the gravitational force on the fluid in the volume V. The first integral on the right side is the momentum flow out through the surface S. The second integral on the right side is the stress force on the surface S. Note that $\sigma_{ij} = \sigma_{ji}$ for a Newtonian fluid. Generally, the integrand in the second integral on the right side will be $\sigma_{ji} \, n_j$. If the partial time derivative in the first term on the left side of Eq. (2.1) is replaced by the material time derivative and if S follows the flow, the first term on the right side disappears. Equations (2.2) and (2.3) require that the fluid is Newtonian. D_{ij} is the deformation rate tensor. μ and λ are called the first and second coefficient of viscosity. Equation (2.2) implies that the average $\frac{1}{3} \sigma_{ii}$ of the normal stress can be written

$$\frac{1}{3} \sigma_{ii} = - \, p + \frac{1}{3} \tau_{ii} \qquad (2.4)$$

Equation (2.3) implies that τ_{ii} can be written

$$\tau_{ii} = (3\lambda + 2\mu) \, v_{i,i} \qquad (2.5)$$

Then, Eq. (2.4) can be written

$$\frac{1}{3} \sigma_{ii} = - \, p + (\lambda + \frac{2}{3} \mu) \, v_{i,i} = - \, p + \kappa \, v_{i,i} \qquad (2.6)$$

$\kappa = \lambda + \frac{2}{3} \mu$ is called the volume viscosity coefficient. Equation (2.3) can be written

$$\tau_{ij} = \kappa \, \delta_{ij} \, v_{k,k} + \mu \, (v_{i,j} + v_{j,i} - \frac{2}{3} \delta_{ij} \, v_{k,k}) \qquad (2.7)$$

The first term on the right side in Eq. (2.7) represents the viscous stresses caused by volume changes. The second term on the right side of Eq. (2.7) represents the viscous stresses caused by shear motions. Stokes' condition means that $\kappa = 0$. Stokes' condition is valid for ideal gases. λ and μ are of the same order of magnitude for liquids but $v_{k,k}$ is generally small since liquids are near incompressible. $v_{k,k} = 0$ for incompressible fluids. This means that it is very common that Eq. (2.7) can be written

$$\tau_{ij} = \mu \, (v_{i,j} + v_{j,i} - \frac{2}{3} \delta_{ij} \, v_{k,k}) \qquad (2.8)$$

The derivative of $\delta_{ij}\, v_{k,k}$ and $\delta_{ij}\, p$ with respect to x_j can be written

$$\left(\delta_{ij}\, v_{k,k}\right)_{,j} = v_{j,ji} \tag{2.9}$$

$$\left(\delta_{ij}\, p\right)_{,j} = p_{,i} \tag{2.10}$$

Then, the derivative of σ_{ij} with respect to x_j can be written

$$\sigma_{ij,j} = -\, p_{,i} + \kappa\, v_{j,ji} + \mu\left(v_{i,jj} + \frac{1}{3} v_{j,ji}\right) \tag{2.11}$$

The Gauss theorem can be written

$$\int_S f_i\, n_i\, dS = \int_V f_{i,i}\, dV \tag{2.12}$$

The Gauss theorem assumes that there are no singularities in the volume V. By applying the Gauss theorem on the surface integrals in Eq. (2.1), the equation can be written

$$\int_V \sigma_{ij,j}\, dV = \int_V \left[\frac{\partial}{\partial t}\left(\rho\, v_i\right) + v_i\left(\rho\, v_j\right)_{,j} + \rho\, v_j\, v_{i,j} - \rho\, g_i\right] dV \tag{2.13}$$

Equation (2.13) is valid for any arbitrary volume V. This means that the integrands in Eq. (2.13) are equal. Since the mass is conserved, the following equation is valid.

$$\int_S \rho\, v_i\, dS = \frac{\partial}{\partial t}\int_V \rho\, dV \tag{2.14}$$

By applying the Gauss theorem on the surface integral in Eq. (2.14), the equation can be written

$$\frac{\partial \rho}{\partial t} + \left(\rho\, v_j\right)_{,j} = 0 \tag{2.15}$$

Equation (2.15) is the continuity equation. Insertion of Eq. (2.15) into Eq. (2.13) gives the following equation.

$$\sigma_{ij,j} = \rho\left(\frac{\partial v_i}{\partial t} + v_j\, v_{i,j} - g_i\right) \tag{2.16}$$

Equation (2.16) is the equation of motion expressed as a differential equation. Insertion of Eq. (2.11) into Eq. (2.16) gives the following equation.

$$\kappa\, v_{j,ji} + \mu\left(v_{i,jj} + \frac{1}{3} v_{j,ji}\right) = \rho\left(\frac{\partial v_i}{\partial t} + v_j\, v_{i,j} - g_i\right) + p_{,i} \tag{2.17}$$

Equation (2.17) is known as the *Navier-Stokes-Duhem equation* for a Newtonian fluid. Stokes' condition means that $\kappa = 0$. Stokes' condition is valid for ideal gases. If the fluid is incompressible, Eq. (2.15) can be written

$$v_{j,j} = 0 \qquad (2.18)$$

Then, Eq. (2.17) can be written

$$\mu \, v_{i,jj} = \rho \left(\frac{\partial v_i}{\partial t} + v_j \, v_{i,j} - g_i \right) + p_{,i} \qquad (2.19)$$

Equation (2.19) is known as the *Navier-Stokes equation* for incompressible flow. Liquids are near incompressible. It is common to use material time derivative in the Navier-Stokes-Duhem equation instead of partial time derivative as in the equations above. The material time derivative $\dot{v}_i$ of v_i can be written

$$\dot{v}_i = \frac{\partial v_i}{\partial t} + v_j \, v_{i,j} \qquad (2.20)$$

where $\dot{v}_i$ is the time derivative of v_i in a point which follows the flow. Note that the velocity vector $\mathbf{v}$ in the equations above is defined in such way that $\rho \, \mathbf{v} = \mathbf{G}$ where $\mathbf{G}$ is the mass flux. There is a definition of the velocity $\mathbf{v}^*$ such that $c \, \mathbf{v}^* = \mathbf{N}$ where c is the total concentration of molecules and $\mathbf{N}$ is the total molar flux. For diffusion in multicomponent systems, the equations above will not be exactly satisfied if $\mathbf{v}^*$ is used instead of $\mathbf{v}$.

2.3   The Bernoulli Equation

If the viscosities μ and κ of the fluid are equal to zero, Eq. (2.17) can be written

$$0 = \rho \left(\frac{\partial v_i}{\partial t} + v_j \, v_{i,j} - g_i \right) + p_{,i} \qquad (2.21)$$

Equation (2.21) is known as the *Euler equation* for an ideal fluid. At stationary flow, the equation can be written

$$0 = \rho \left(v_j \, v_{i,j} - g_i \right) + p_{,i} \qquad (2.22)$$

At stationary flow where the velocity is zero everywhere, the equation can be written

$$0 = - \rho \, g_i + p_{,i} \qquad (2.23)$$

Equation (2.23) describes the hydrostatic equilibrium. In a compressible fluid, there is a connection between the pressure p, the density ρ and the absolute temperature

T. The ideal gas law is a well-known example of such a connection. In a *barotropic fluid*, there is no temperature in the state relationship. In this case, the density ρ is only a function of the pressure p. In a barotropic fluid, the pressure function $P(p)$ can be defined

$$P(p) = \int_{p_0}^{p} \frac{dp}{\rho} \tag{2.24}$$

Since the gravitational field g_i is conservative, g_i can be written

$$g_i = -\Omega_{,i} \quad \text{or} \quad \mathbf{g} = -\nabla\Omega \tag{2.25}$$

where Ω is the potential energy. Then, Eq. (2.23) can be written

$$0 = (\Omega + P)_{,i} \tag{2.26}$$

For a barotropic fluid, Eq. (2.21) can be written

$$0 = \frac{\partial v_i}{\partial t} + v_j v_{i,j} + (\Omega + P)_{,i} \tag{2.27}$$

Integration of Eq. (2.27) with respect to x_i gives

$$\int \frac{\partial v_i}{\partial t} dx_i + \int v_j v_{i,j} dx_i + \Omega + P = C(t) \tag{2.28}$$

Along a streamline, the equation $dx_i = (v_i/v) ds$ is valid where ds is an infinitesimal displacement along the streamline and v is the absolute value of the velocity vector. Then, $v_j v_{i,j} dx_i$ in the second integral in Eq. (2.28) can be written

$$v_j v_{i,j} dx_i = v_j v_{i,j} (v_i/v) ds = v_i v_{i,j} (v_j/v) ds = v_i v_{i,j} dx_j = v_i dv_i \tag{2.29}$$

Then, the second integral in Eq. (2.28) can be written

$$\int v_j v_{i,j} dx_i = \int v_i dv_i = \frac{1}{2} v_i v_i = \frac{1}{2} v^2 \tag{2.30}$$

Then, Eq. (2.28) can be written

$$\int \frac{\partial v_i}{\partial t} dx_i + \frac{1}{2} v^2 + \Omega + P = C(t) \tag{2.31}$$

Equation (2.31) is known as the *Bernoulli equation*. Note that $C(t)$ is constant along a streamline but the value of $C(t)$ can be different for different streamlines. If the flow is rotationless ($\nabla \times \mathbf{v} = 0$), C can be assumed to be time independent and C is equal for all streamlines. (see Sect. 2.4). At stationary flow, Eq. (2.31) can be written

$$\frac{1}{2} v^2 + \Omega + P = C \tag{2.32}$$

For incompressible flow, Eq. (2.32) can be written

$$\frac{1}{2} v^2 + \frac{p}{\rho} + \Omega = C \tag{2.33}$$

If the gravitational field is homogeneous, Eq. (2.33) can be written

$$\frac{1}{2} v^2 + \frac{p}{\rho} + g h = C \tag{2.34}$$

where g is the absolute value of the vector (g_1, g_2, g_3) and h is the length coordinate in opposite direction to the gravitational field. If the internal energy in a gas is proportional to the temperature, the gas is called *polytropic*. For ideal gases and polytropic gases, the barotropic state relationship is valid

$$p = p_0 \left(\frac{\rho}{\rho_0} \right)^\gamma \tag{2.35}$$

$$\gamma = \frac{c_p}{c_v} \tag{2.36}$$

where ρ is the density of the gas at pressure p and ρ_0 is the density of the gas at pressure p_0. Then, the pressure function (2.24) can be written

$$P = \int_{p_0}^{p} \frac{dp}{\rho} = \frac{\gamma}{\gamma - 1} \left(\frac{p}{\rho} - \frac{p_0}{\rho_0} \right) \tag{2.37}$$

Assume that $\Omega = 0$ in Eq. (2.32). Insertion of Eq. (2.37) into Eq. (2.32) means that Eq. (2.32) can be written

$$\frac{1}{2} v^2 + \frac{\gamma}{\gamma - 1} \frac{p}{\rho} = \frac{1}{2} u^2 \tag{2.38}$$

where u is a constant. Furthermore, the following equation is valid.

$$\lim_{\rho \to 0} \frac{p}{\rho} = 0 \tag{2.39}$$

Equation (2.39) means that u is the theoretical maximum speed of the flow. The speed of sound c is obtained from the following equation.

$$c^2 = \frac{dp}{d\rho} = \gamma \frac{p}{\rho} \tag{2.40}$$

Insertion of Eq. (2.40) into Eq. (2.38) means that Eq. (2.38) can be written

$$\frac{1}{2}v^2 + \frac{c^2}{\gamma - 1} = \frac{1}{2}u^2 \tag{2.41}$$

Equation (2.41) is called the Bernoulli equation for a polytropic gas.

2.4 Potential Flow

Rotationless flow means that the line integral of the velocity field along any closed curve C in a fluid is equal to zero.

$$\oint_C v_i \, dx_i = 0 \tag{2.42}$$

The Stokes theorem can be written

$$\oint_C f_i \, dx_i = \int_S \epsilon_{ijk} \, f_{k,j} \, n_i \, dS = \int_S (\nabla \times \mathbf{f}) \cdot d\mathbf{S} \tag{2.43}$$

where S is a surface enclosed by the closed curve C. Since Eq. (2.43) is valid for any arbitrary closed curve C, the integrand in the surface integral is equal to zero if the surface integral is equal to zero. This means that the following equation is valid for rotationless flow.

$$\epsilon_{ijk} \, v_{k,j} = 0 \text{ or } \nabla \times \mathbf{v} = 0 \tag{2.44}$$

$e_{ijk} \, v_{k,j}$ or $\nabla \times \mathbf{v}$ is the curl of the velocity field. Equation (2.27) is valid for an ideal barotropic fluid in a conservative gravitational field. If the flow is rotationless, the velocity vector v_i can be written

$$v_i = -\Phi_{,i} \text{ or } \mathbf{v} = -\nabla \Phi \tag{2.45}$$

One can show that $\nabla \times \nabla \Phi = 0$. Φ is a scalar function which is called the *velocity potential* of the flow. Note that Φ sometimes is defined in such a way that $\mathbf{v} = \nabla \Phi$ instead of $\mathbf{v} = -\nabla \Phi$. Equation (2.45) is always valid in this book. The continuity equation Eq. (2.15) can be written

$$\frac{\partial \rho}{\partial t} - \left(\rho \, \Phi_{,i} \right)_{,i} = 0 \tag{2.46}$$

Equation (2.27) can be written

$$0 = -\frac{\partial}{\partial t} \Phi_{,i} + \Phi_{,j} \, \Phi_{,ij} + (\Omega + P)_{,i} \tag{2.47}$$

The second term in (2.47) can be written

$$\Phi_{,j}\,\Phi_{,ij} = \Phi_{,j}\,\Phi_{,ji} = \frac{1}{2}\left(\Phi_{,j}^2\right)_{,i} = \frac{1}{2}\left(v^2\right)_{,i} \tag{2.48}$$

Then, Eq. (2.47) can be written

$$0 = \left(-\frac{\partial\Phi}{\partial t} + \frac{1}{2}v^2 + \Omega + P\right)_{,i} \tag{2.49}$$

Integration of (2.49) with respect to x_i gives the following equation.

$$-\frac{\partial\Phi}{\partial t} + \frac{1}{2}v^2 + \Omega + P = C \tag{2.50}$$

Equation (2.50) is called the *pressure equation*. For stationary flow, Eq. (2.50) is identical to Eq. (2.32). For rotationless flow, Eq. (2.31) is identical to Eq. (2.50). The constant C in Eq. (2.50) is independent of the space coordinates x_i. The velocity potential Φ can be defined in such a way that C even is independent of the time t since $\mathbf{v} = -\nabla\Phi$. The velocity vector $\mathbf{v}$ is not influenced by how C varies in time. This means, for example, that C can be set to zero. If the flow is not rotationless, the velocity vector $\mathbf{v}$ depends on how C varies in time. The Eqs. (2.31) and (2.50) are similar but not identical. For stationary and rotationless flow of an ideal incompressible fluid, the continuity equation Eq. (2.46) can be written

$$\Phi_{,ii} = 0 \quad\text{or}\quad \nabla^2\Phi = 0 \tag{2.51}$$

Equation (2.51) is the Laplace equation. If the internal energy in a gas is proportional to the temperature, the gas is called *polytropic*. For stationary flow of a polytropic gas, Eq. (2.47) can be written

$$v_j\,\Phi_{,ij} = \frac{1}{\rho}\,p_{,i} \tag{2.52}$$

where $v_j = -\Phi_{,j}$. The continuity equation Eq. (2.46) can be written

$$v_i\,\rho_{,i} = \rho\,\Phi_{,ii} \tag{2.53}$$

Furthermore, the following equation is valid

$$p_{,i} = \frac{dp}{d\rho}\,\rho_{,i} = c^2\,\rho_{,i} \tag{2.54}$$

where c is the speed of sound. Multiplying Eq. (2.52) by v_i and inserting Eq. (2.54) and (2.53) into Eq. (2.52) gives

$$v_i \, v_j \, \Phi_{,ij} \;=\; \frac{c^2}{\rho} \, v_i \, \rho_{,i} \;=\; \frac{c^2}{\rho} \, \rho \, \Phi_{,ii} \qquad \text{or} \tag{2.55}$$

$$\Phi_{,ii} \;-\; \frac{v_i \, v_j}{c^2} \, \Phi_{,ij} \;=\; 0 \tag{2.56}$$

Equation (2.56) is called the *gas dynamic equation*. The equation is called *quasi stationary*. The equation is elliptic if the flow is subsonic and hyperbolic if the flow is supersonic.

2.5 The Wave Equation

A soundwave is an oscillating motion with small amplitude in a compressible medium. The medium is assumed to be in thermodynamic equilibrium and the motion occurs isentropically. This section describes longitudinal sound propagation in a homogeneous fluid at equilibrium at constant pressure p_0 and constant density ρ_0. The pressure p and the density ρ can be written

$$p = p_0 + \Delta p \tag{2.57}$$

$$\rho = \rho_0 + \Delta \rho \tag{2.58}$$

where $\Delta p / p_0 \ll 1$ and $\Delta \rho / \rho_0 \ll 1$. For small fluctuations, the second term in Eq. (2.50) can be neglected. If Ω in Eq. (2.50) is neglected, the pressure equation can approximately be written

$$\frac{\partial \Phi}{\partial t} \approx \int_{p_0}^{p_0 + \Delta p} \frac{dp}{\rho} \approx \frac{\Delta p}{\rho_0} \tag{2.59}$$

The continuity equation Eq. (2.46) can approximately be written

$$\frac{\partial \Delta \rho}{\partial t} \approx \rho_0 \, \Phi_{,ii} \tag{2.60}$$

Derivation of Eq. (2.59) with respect to t gives

$$\frac{\partial^2 \Phi}{\partial t^2} = \frac{1}{\rho_0} \frac{\partial \Delta p}{\partial t} = \frac{1}{\rho_0} \frac{dp}{d\rho} \frac{\partial \Delta \rho}{\partial t} = \frac{1}{\rho_0} c^2 \frac{\partial \Delta \rho}{\partial t} \tag{2.61}$$

Insertion of Eq. (2.60) into Eq. (2.61) gives

$$\Phi_{,ii} = \frac{1}{c^2} \frac{\partial^2 \Phi}{\partial t^2} \qquad \text{or} \qquad \nabla^2 \Phi = \frac{1}{c^2} \frac{\partial^2 \Phi}{\partial t^2} \tag{2.62}$$

Equation (2.62) is known as the wave equation for soundwaves in a compressible fluid. For a given velocity potential Φ, the velocity **v** and the sound pressure Δp are obtained from the following equations.

$$v_i = - \Phi_{,i} \quad \text{or} \quad \mathbf{v} = - \nabla \Phi \qquad (2.63)$$

$$\Delta p = \rho_0 \frac{\partial \Phi}{\partial t} \qquad (2.64)$$

The barotropic state relationship for a polytropic gas can be written

$$p = p_0 \left(\frac{\rho}{\rho_0} \right)^\gamma \qquad (2.65)$$

$$\gamma = \frac{c_p}{c_v} \qquad (2.66)$$

where ρ is the density of the gas at pressure p and ρ_0 is the density of the gas at pressure p_0. The value of γ is 1.40 for air. The speed of sound c can be calculated from the following equation.

$$c^2 = \gamma \frac{p_0}{\rho_0} \qquad (2.67)$$

The value of the speed of sound in air is 346 m/s at pressure 1 bar and temperature 25 °C.

2.6 Friction Factor

Consider a fluid with constant viscosity μ and density ρ. For such a fluid, Eq. (2.19) is valid. The fluid is incompressible. Assume that the gravitational field $g_i = 0$ and that the flow is stationary and laminar. Then, Eq. (2.19) can be written

$$\mu \, v_{i,jj} = \rho \, v_j \, v_{i,j} + p_{,i} \qquad (2.68)$$

Assume that the fluid flows around a solid object with a given geometry. For uniform objects, the geometry of the object can be specified with only one length quantity D. D can, for example, be the diameter of a sphere or the diameter of a cylinder with constant ratio between diameter and length. Further, assume that the object is oriented in a given way relative to the flow direction and that the flow at infinite distance from the object is homogeneous with constant velocity v_0. In this case, it is possible to introduce dimensionless space coordinates ξ_i, dimensionless velocities v_i' and dimensionless pressure p' in the following way.

$$\xi_i = \frac{x_i}{D} \qquad (2.69)$$

$$v_i' = \frac{v_i}{v_0} \tag{2.70}$$

$$p' = \frac{p}{\rho v_0^2} \tag{2.71}$$

Then, Eq. (2.68) can be written

$$\frac{\partial^2 v_i'}{\partial \xi_1^2} + \frac{\partial^2 v_i'}{\partial \xi_2^2} + \frac{\partial^2 v_i'}{\partial \xi_3^2}$$
$$= \frac{\rho v_0 D}{\mu} \left(v_1' \frac{\partial v_i'}{\partial \xi_1} + v_2' \frac{\partial v_i'}{\partial \xi_2} + v_3' \frac{\partial v_i'}{\partial \xi_3} + \frac{\partial p'}{\partial \xi_i} \right) \tag{2.72}$$

The only parameter in Eq. (2.72) is $\rho v_0 D/\mu$. This parameter is known as the *Reynolds number Re*.

$$Re = \frac{\rho v_0 D}{\mu} \tag{2.73}$$

The solution (v_1', v_2', v_3', p') to Eq. (2.72) in the variable space (ξ_1, ξ_2, ξ_3) for a given shape of the object depends only on Re. This means that the flow pattern for uniform objects is the same for all values of D for constant value of Re. The total force F_i on the object can be written

$$F_i = \int_S \sigma_{ji} n_j \, dS = \int_S \left[-p \delta_{ij} + \mu \left(v_{i,j} + v_{j,i} \right) \right] n_j \, dS$$
$$= \rho v_0^2 D^2 \int_{S'} \left[-p' \delta_{ij} + \frac{1}{Re} \left(\frac{\partial v_i'}{\partial \xi_j} + \frac{\partial v_j'}{\partial \xi_i} \right) \right] n_j \, dS' \tag{2.74}$$

where S is the surface of the object, S' is the surface of the object in the variable space (ξ_1, ξ_2, ξ_3) and $dS = D^2 dS'$. The last integral in Eq. (2.74) is a dimensionless function which only depends on the Reynolds number Re. Assume that the flow is directed in the x_1 coordinate direction and the geometry of the object is such that $F_2 = F_3 = 0$. The projected area A of the object in the flow direction is proportional to D^2. Then, F_1 can be written

$$F_1 = F = \frac{1}{2} \rho v_0^2 A f(Re) \tag{2.75}$$

The function $f(Re)$ is defined as the *friction factor* of the object. Note that it is not entirely clear how to define the Reynolds number Re of an object. For an object with an irregular geometry, the meaning of the size D may be uncertain. The friction factor $f(Re)$ depends on how D is defined.

2.7 Analytical Solutions to the Navier-Stokes Equation

This section contains analytical solutions to the Navier-Stokes equation in some simple geometries. It is very difficult to find analytical solutions to the Navier-Stokes equation if one can not assume that either the frictional forces or the acceleration forces are zero or negligible even for very simple geometries. An example of this is radial flow in a slit. Most analytical solutions in this section apply in geometries where the acceleration forces are zero. One exception is Couette flow between two rotating cylinders.

2.7.1 Flow Around a Sphere

Consider a sphere of diameter D and radius R in a flowing incompressible fluid where the flow is stationary and homogeneous with velocity v_0 at infinite distance from the sphere. Assume that the flow is directed in the z-axis direction and that the center of the sphere is placed at the origin in Fig. 2.1.

This problem can be solved analytically for creeping flow. Creeping flow means that the flow rate is so low that the frictional forces are dominant compared to the acceleration forces. This means that Eq. (2.68) can be written

$$\mu\, v_{i,jj} = p_{,i} \tag{2.76}$$

The solution is rotationally symmetric around the z-axis, which means that the solution only depends on the coordinates r and θ and that $v_\varphi = 0$. In spherical coordinates, Eq. (2.76) can be written

$$\mu \left(\nabla^2 v_r - \frac{2\, v_r}{r^2} - \frac{2}{r^2}\frac{\partial v_\theta}{\partial \theta} - \frac{2\, v_\theta \cot \theta}{r^2} \right) = \frac{\partial p}{\partial r} \tag{2.77}$$

$$\mu \left(\nabla^2 v_\theta + \frac{2}{r^2}\frac{\partial v_r}{\partial \theta} - \frac{v_\theta}{r^2 \sin^2 \theta} \right) = \frac{1}{r}\frac{\partial p}{\partial \theta} \tag{2.78}$$

Fig. 2.1 Spherical coordinates

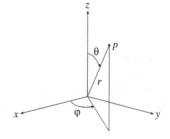

$$\nabla^2 = \frac{1}{r^2} \frac{\partial}{\partial r} \left(r^2 \frac{\partial}{\partial r} \right) + \frac{1}{r^2 \sin \theta} \frac{\partial}{\partial \theta} \left(\sin \theta \frac{\partial}{\partial \theta} \right) \tag{2.79}$$

The solution can be written

$$v_r = v_0 \left[1 - \frac{3}{2} \left(\frac{R}{r} \right) + \frac{1}{2} \left(\frac{R}{r} \right)^3 \right] \cos \theta \tag{2.80}$$

$$- v_\theta = v_0 \left[1 - \frac{3}{4} \left(\frac{R}{r} \right) - \frac{1}{4} \left(\frac{R}{r} \right)^3 \right] \sin \theta \tag{2.81}$$

$$p = p_0 - \frac{3}{2} \frac{\mu v_0}{R} \left(\frac{R}{r} \right)^2 \cos \theta \tag{2.82}$$

where p_0 is the pressure at infinite distance from the sphere. This solution is valid with high accuracy when the Reynolds number $D v_0 \rho / \mu$ is less than about 0.1. Then, the shear stress $\tau_{r\theta} = \tau_{\theta r}$ can be written

$$\tau_{r\theta} = - \mu \left[r \frac{\partial}{\partial r} \left(\frac{v_\theta}{r} \right) + \frac{1}{r} \frac{\partial v_r}{\partial \theta} \right] = \frac{3}{2} \frac{\mu v_0}{r} \left(\frac{R}{r} \right)^3 \sin \theta \tag{2.83}$$

The total force F_i on the sphere can be written

$$F_i = \int_S \sigma_{ji} n_j \, dS = \int_S \left(- p \, \delta_{ij} + \tau_{ij} \right) n_j \, dS \tag{2.84}$$

where S is the surface of the sphere. Equations (2.82) and (2.83) means that the force F_z on the sphere in the z-axis direction can be written

$$F_z = 3\pi \mu R v_0 \int_0^\pi \left(\sin^2 \theta + \cos^2 \theta - \frac{2}{3} \frac{p_0 R}{\mu v_0} \sin \theta \cos \theta \right) \sin \theta \, d\theta$$

$$= 3\pi \mu R v_0 \int_0^\pi \sin \theta \, d\theta = 6\pi \mu R v_0 \tag{2.85}$$

Equation (2.85) is known as the *Stokes law*. The Stokes law is, in particular, useful in the theory of sedimentation. Note that the formula only applies to Reynolds numbers which are less than about 0.1. At $Re = 1$, the formula gives a value which is about 10 % too low. The Stokes law can also be written

$$F = \frac{1}{2} \rho v_0^2 A f(Re) \tag{2.86}$$

$$f(Re) = \frac{24}{Re} \tag{2.87}$$

$$Re = \frac{\rho \, v_0 \, D}{\mu} \tag{2.88}$$

$$A = \pi R^2 \tag{2.89}$$

The function $f(Re) = 24/Re$ is called the Stokes friction factor. There are no exact analytical expressions for the friction factor at high Reynolds numbers. In the transition region, the friction factor can approximately be written

$$f(Re) = \frac{18.5}{Re^{3/5}} \quad 2 < Re < 500 \tag{2.90}$$

The transition region is the region where the friction forces are of the same order of magnitude as the acceleration forces. At higher values of Re, the acceleration forces dominate compared with the friction forces and the friction factor is approximately independent of Re.

$$f(Re) = 0.44 \quad 500 < Re < 2 \cdot 10^5 \tag{2.91}$$

When the friction factor is a constant, Eq. (2.86) is called Newton's law for the friction force on a sphere.

2.7.2   Flow in a Cylindrical Tube

Consider a cylindrical channel with diameter D and radius R containing a flowing incompressible fluid with constant viscosity μ. Assume that the channel center line coincides with the z-axis in a cylindrical coordinate system as in Fig. 2.2.

Assume that the flow is stationary and independent of the coordinates z and θ. This applies to fully developed flow profile. Further, assume that all points which follow the flow, move with constant velocity along straight lines. Then, the first term on the right side of Eq. (2.68) is equal to zero. In cylindrical coordinates, Eq. (2.68) can be written

$$\mu \, \frac{1}{r} \frac{\partial}{\partial r} \left(r \, \frac{\partial v_z}{\partial r} \right) = \frac{\partial p}{\partial z} \tag{2.92}$$

Fig. 2.2 Cylindrical coordinates

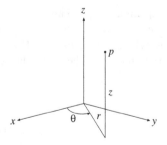

The solution $v_z(r)$ can be written

$$v_z(r) = v_0 \left(1 - \frac{r^2}{R^2} \right) \tag{2.93}$$

where v_0 is the velocity in the middle of the cylinder. The shear stress $\tau_{zr} = \tau_{rz}$ can be written

$$\tau_{zr} = -\mu \frac{\partial v_z}{\partial r} = \frac{2 r \mu v_0}{R^2} \tag{2.94}$$

Consider a segment of the cylinder surface which is located between the coordinates z_1 and z_2. This surface is denoted ΔS. The force ΔF_z on this surface in the z-coordinate direction can be written

$$\Delta F_z = \int_{\Delta S} \sigma_{ji} n_j \, dS = 4 \pi \mu v_0 \Delta z \tag{2.95}$$

where $\Delta z = z_2 - z_1$. If S_z is the cross-sectional surface of the channel, the volume flow q through the channel can be written

$$q = \int_{S_z} v_i n_i \, dS = 2 \pi \int_0^R v_z r \, dr = \frac{\pi}{2} v_0 R^2 \tag{2.96}$$

Then, the force ΔF_z can be written

$$\Delta F_z = \frac{8 \mu q}{R^2} \Delta z \tag{2.97}$$

The pressure difference Δp_z can be written

$$\Delta p_z = \frac{8 \mu q}{\pi R^4} \Delta z \tag{2.98}$$

Equation (2.98) is known as the *Hagen-Poiseuille law*. Note that $\partial p / \partial z$ in Eq. (2.92) is a constant which is independent of the coordinates r and z and have the value $-\Delta p_z / \Delta z$. Then, Eq. (2.98) is obtained from Eqs. (2.92), (2.93) and (2.96) without the need to calculate the shear force on the tube wall. The linear velocity $\langle v \rangle$ is defined

$$\langle v \rangle = \frac{q}{\pi R^2} \tag{2.99}$$

The friction factor $f(Re)$ for a tube with inner diameter D is defined

$$\Delta p_z = \frac{1}{2} \rho \langle v \rangle^2 \frac{4 \Delta z}{D} f(Re) \tag{2.100}$$

The Reynolds number Re is defined

$$Re = \frac{\rho \langle v \rangle D}{\mu} \tag{2.101}$$

Equations (2.98) and (2.100) means that the friction factor $f(Re)$ can be written

$$f(Re) = \frac{16}{Re} \tag{2.102}$$

Equation (2.102) applies to Reynolds numbers which are less than about 2100. For higher values of Re, the flow becomes unstable and turns into turbulent flow. For undisturbed laminar flow, there are no acceleration forces in the flow and the first term on the right side of Eq. (2.68) is zero. Disturbances in the flow give rise to acceleration forces which increase faster with increasing value of Re than the friction forces which are represented by the left side of Eq. (2.68). For sufficiently high values of Re, these disturbances increase spontaneously and the flow becomes chaotic. For turbulent flow, the friction factor can approximately be written

$$f(Re) = \frac{0.0791}{Re^{1/4}} \qquad 2.1 \cdot 10^3 < Re < 10^5 \tag{2.103}$$

Equation (2.103) is called the *Blasius formula*. The formula is experimentally determined and is valid for smooth pipes.

2.7.3 Flow in a Rectangular Channel

Consider a rectangular channel with the width $2a$ and the height $2b$ containing a flowing incompressible fluid with constant viscosity μ. Assume that the channel center line coincides with the z-axis in a Cartesian coordinate system as in Fig. 2.3.

Assume that the flow is stationary and independent of the coordinate z. This applies to fully developed flow profile. Further, assume that all points which follow the flow, move with constant velocity along straight lines. Then, the first term on the

Fig. 2.3 Rectangular channel

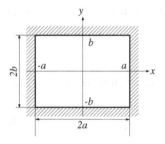

right side of Eq. (2.68) is zero, since there are no acceleration forces in the flow. In Cartesian coordinates, Eq. (2.68) can be written

$$\mu \left(\frac{\partial^2 v_z}{\partial x^2} + \frac{\partial^2 v_z}{\partial y^2} \right) = \frac{\partial p}{\partial z} \qquad (2.104)$$

$\partial p / \partial z$ is, in this case, a constant independent of x, y and z. Then, Eq. (2.104) can be written

$$\frac{\partial^2 v_z}{\partial x^2} + \frac{\partial^2 v_z}{\partial y^2} = - C = - \frac{\Delta p}{\mu \, \Delta z} \qquad (2.105)$$

where Δp is the pressure difference between two points with the distance Δz along the z-axis. C is a positive constant. Equation (2.105) is the Poisson equation with constant source term. Assume that v_z can be written as the sum of a series of orthogonal functions in the following way.

$$v_z(x, y) = \sum_n \sum_m v_{nm} \, f_n(x) \, f_m(y) \qquad (2.106)$$

Then, the constant C can be written

$$C = \sum_n \sum_m c_{nm} \, f_n(x) \, f_m(y) \qquad (2.107)$$

where the sum in Eq. (2.107) is Fourier series or orthogonal series to the constant C in the function system $f_n(x) \, f_m(y)$. The Fourier coefficients c_{nm} can be written

$$c_{nm} = C \frac{\int\limits_{-a}^{a} f_n(x) \, dx \int\limits_{-b}^{b} f_m(y) \, dy}{\int\limits_{-a}^{a} f_n^2(x) \, dx \int\limits_{-b}^{b} f_m^2(y) \, dy} \qquad (2.108)$$

If Eqs. (2.107) and (2.109) are satisfied, Eq. (2.105) is also satisfied.

$$v_{nm} \left[f_n^{''}(x) \, f_m(y) + f_n(x) \, f_m^{''}(y) \right] = - c_{nm} \, f_n(x) \, f_m(y) \qquad (2.109)$$

Division of Eq. (2.109) by $v_{nm} f_n(x) f_m(y)$ gives the following equation.

$$\frac{f_n^{''}(x)}{f_n(x)} + \frac{f_m^{''}(y)}{f_m(y)} = - \frac{c_{nm}}{v_{nm}} \qquad (2.110)$$

The right side of Eq. (2.110) is a constant, independent of x and y. This means that the both terms on the left side are constants. Assume that the values of these terms are $-\lambda_n^2$ and $-\lambda_m^2$. This means that the following equations are valid.

$$f_n''(x) + \lambda_n^2 f_n(x) = 0 \tag{2.111}$$

$$f_m''(y) + \lambda_m^2 f_m(y) = 0 \tag{2.112}$$

The solution to Eqs. (2.111) and (2.112) can be written

$$f_n(x) = A_n \cos(\lambda_n x) + B_n \sin(\lambda_n x) \tag{2.113}$$

$$f_m(y) = A_m \cos(\lambda_m y) + B_m \sin(\lambda_m y) \tag{2.114}$$

The solution is symmetric around the z-axis. This means that $f_n(x) = f_n(-x)$ and $f_m(y) = f_m(-y)$ and that B_n and B_m are equal to zero. The velocity v_z is equal to zero at the channel walls. This means that $\lambda_n a = \pi/2 + n\pi$ and $\lambda_m b = \pi/2 + m\pi$ where n and m are integers. Then, λ_n and λ_m can be written.

$$\lambda_n = \frac{\pi}{2a}(1 + 2n) \qquad n \in \{0, 1, 2, 3, ..., \infty\} \tag{2.115}$$

$$\lambda_m = \frac{\pi}{2b}(1 + 2m) \qquad m \in \{0, 1, 2, 3, ..., \infty\} \tag{2.116}$$

Assume that $A_n = A_m = 1$. Then, $f_n(x)$ and $f_m(y)$ can be written

$$f_n(x) = \cos(\lambda_n x) \tag{2.117}$$

$$f_m(y) = \cos(\lambda_m y) \tag{2.118}$$

This function system is a complete orthogonal function system. Equation (2.108) implies that c_{nm} can be written

$$c_{nm} = \frac{16C(-1)^{n+m}}{\pi^2(1+2n)(1+2m)} \tag{2.119}$$

One can show that Eq. (2.119) implies that Eq. (2.107) is satisfied. This means that Eq. (2.106) is a solution to Eq. (2.105) with the current boundary conditions. Equations (2.110) and (2.119) means that v_{nm} can be written

$$v_{nm} = \frac{64\,ab\,C\,(-1)^{n+m}}{\pi^4(1+2n)(1+2m)\left[\dfrac{b}{a}(1+2n)^2 + \dfrac{a}{b}(1+2m)^2\right]} \tag{2.120}$$

The solution $v_z(x, y)$ can be written

$$v_z(x, y) = \sum_{n=0}^{\infty}\sum_{m=0}^{\infty} v_{nm} \cos\left[\frac{\pi x}{2a}(1+2n)\right]\cos\left[\frac{\pi y}{2b}(1+2m)\right] \tag{2.121}$$

The volume flow q in the channel can be written

$$q = \int\limits_{-a}^{a} \int\limits_{-b}^{b} v(x, y)\, dxdy = 16 \left(\frac{2}{\pi}\right)^6 a^2 b^2 \frac{\Delta p}{\mu\, \Delta z} \Sigma \tag{2.122}$$

$$\Sigma = \sum_{n=0}^{\infty} \sum_{m=0}^{\infty} \frac{1}{(1+2n)^2 (1+2m)^2 \left[\dfrac{b}{a}(1+2n)^2 + \dfrac{a}{b}(1+2m)^2\right]} \tag{2.123}$$

The pressure drop Δp for the flow q can be calculated from Eq. (2.122). Table 2.1 shows values of Σ for different values of a/b. The hydraulic diameter D_h of a rectangular channel with the width $2a$ and the height $2b$ is defined

$$D_h = \frac{4\,ab}{a+b} \tag{2.124}$$

The linear velocity $\langle v \rangle$ is defined

$$\langle v \rangle = \frac{q}{4\,ab} \tag{2.125}$$

The Reynolds number Re is defined

$$Re = \frac{\rho \langle v \rangle D_h}{\mu} \tag{2.126}$$

The friction factor $f(Re)$ is defined

$$\Delta p = \frac{1}{2} \rho \langle v \rangle^2 \frac{4\Delta z}{D_h} f(Re) \tag{2.127}$$

Then, $f(Re)$ can be written

$$f(Re) = \frac{C_f}{Re} \tag{2.128}$$

$$C_f = \frac{2\left(\dfrac{a}{b}\right)}{\left(\dfrac{2}{\pi}\right)^6 \left(1 + \dfrac{a}{b}\right)^2 \Sigma} \tag{2.129}$$

Table 2.1 shows values of C_f for different values of a/b. Note that the value of C_f is equal to 16 for a cylindrical channel with the diameter D_h. Equation (2.128) is only valid for laminar flow. The following formula gives a diameter which approximately gives the same pressure drop in a cylindrical tube with diameter D as in a rectangular channel with the sides $2a$ and $2b$.

Table 2.1 Σ and C_f as functions of a/b

a/b	Σ	C_f
1.0	0.528	14.2
0.9	0.525	14.3
0.8	0.516	14.4
0.7	0.498	14.6
0.6	0.470	15.0
0.5	0.429	15.5
0.4	0.375	16.4
0.3	0.305	17.5
0.2	0.219	19.1
0.1	0.117	21.2
0.0	0.000	24.0

$$D = 2.60 \frac{(ab)^{0.625}}{(a+b)^{0.250}} \tag{2.130}$$

In a channel with the sides a and b, the coefficient 2.60 is replaced by 1.30 in Eq. (2.130).

2.7.4 Linear Flow in a Slit

Consider a slit with the width $2a$ containing a flowing incompressible fluid with constant viscosity μ. Assume that the center of the slit coincides with the y-z-plane in a Cartesian coordinate system as in Fig. 2.4 and that the slit has infinite extension in the y-z-plane.

Assume that the flow is stationary and independent of the coordinates y and z and that the flow is directed in the z coordinate direction. The flow pattern is independent of z for fully developed flow profile. Further, assume that all points which follow the flow, move with constant velocity along straight lines. Then, the first term on the right side of (2.68) is equal to zero, since there are no acceleration forces in the flow. Then, in Cartesian coordinates, Eq. (2.68) can be written

Fig. 2.4 Infinite slit

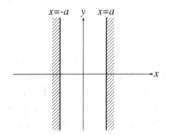

$$\mu \frac{\partial^2 v_z}{\partial x^2} = \frac{\partial p}{\partial z} \qquad (2.131)$$

$\partial p / \partial z$ is, in this case, a constant, independent of x, y and z. Then, Eq. (2.131) can be written

$$\frac{\partial^2 v_z}{\partial x^2} = -C = -\frac{\Delta p}{\mu \Delta z} \qquad (2.132)$$

where Δp is the pressure difference between two points with the distance Δz along the z-axis. C is a positive constant. Equation (2.132) is the Poisson equation with constant source term. The solution can be written

$$v_z(x) = v_0 \left(1 - \frac{x^2}{a^2} \right) \qquad (2.133)$$

$$v_0 = \frac{a^2 \Delta p}{2 \mu \Delta z} \qquad (2.134)$$

where v_0 is the velocity in the middle of the slit. The linear velocity $\langle v \rangle$ is defined

$$\langle v \rangle = \frac{1}{2a} \int_{-a}^{a} v_z(x) \, dx = \frac{2}{3} v_0 = \frac{a^2 \Delta p}{3 \mu \Delta z} \qquad (2.135)$$

The hydraulic diameter D_h of a slit with the width $2a$ is defined

$$D_h = 4a \qquad (2.136)$$

Then, the friction factor $f(Re)$ can be written

$$f(Re) = \frac{24}{Re} \qquad (2.137)$$

$$Re = \frac{\rho \langle v \rangle D_h}{\mu} \qquad (2.138)$$

Equation (2.137) is only valid for laminar flow.

2.7.5 Radial Flow in a Slit

Consider a slit with the width $2a$ containing a flowing incompressible fluid with constant viscosity μ. Assume that the center of the slit coincides with the plane $(r, \theta, 0)$ in a cylindrical coordinate system as in Fig. 2.5 and that the flow is radially directed with fully developed flow profile. See also Fig. 2.2.

Fig. 2.5 Radial flow in a slit

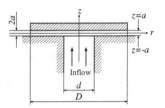

The inflow occurs through a cylindrical hole with the diameter d in the slit wall. The center line of the cylinder coincides with the z-axis. A circular disc of diameter D is located at the distance $2a$ from the wall. The z-axis is passing through the center point of the disc. At sufficiently large values of the coordinate r, one can assume that flow profile is fully developed. Assume that the flow is stationary and independent of the angle coordinate θ. In cylindrical coordinates, Eq. (2.68) can be written

$$\mu \left[\frac{1}{r} \frac{\partial}{\partial r} \left(r \frac{\partial v_r}{\partial r} \right) + \frac{\partial^2 v_r}{\partial z^2} \right] = \rho\, v_r \frac{\partial v_r}{\partial r} + \frac{\partial p}{\partial r} \tag{2.139}$$

The right side in Eq. (2.139) can be written

$$\rho\, v_r \frac{\partial v_r}{\partial r} + \frac{\partial p}{\partial r} = \frac{\partial}{\partial r} \left(\frac{1}{2} \rho\, v_r^2 + p \right) \tag{2.140}$$

For frictionless flow, the left side of Eq. (2.139) is equal to zero which leads to the following equation where C is a constant, independent of r.

$$\frac{1}{2} \rho\, v_r^2 + p = C \tag{2.141}$$

Equation (2.141) is the Bernoulli equation for stationary incompressible flow when the gravitational field is equal to zero. If v_r is independent of z, the flow is rotationless. Then, C is independent of z. Note that the pressure increases with increasing r in this case. The velocity v_r can be written

$$v_r = \frac{q}{4\pi\, a r} \tag{2.142}$$

where q is the volume flow in the slit. Assume that p_0 is the pressure outside the disc in Fig. 2.5. The velocity of the fluid is approximately zero in this region. This means that $C \approx p_0$. In the region $0 < r < d/2$ inside the disc, the velocity of the fluid is also approximately zero. p is therefore approximately equal to p_0 even in this region. In the region $d/2 < r < D/2$, the pressure can approximately be written

$$p \approx p_0 - \frac{\rho\, q^2}{32\, \pi^2 a^2 r^2} \tag{2.143}$$

Then, the total force F on the disc in the z-axis direction can approximately be written

$$F \approx -\frac{\rho q^2}{16\pi a^2} \ln \frac{D}{d} \tag{2.144}$$

In this case, the force on the disc in Fig. 2.5 is directed towards the hole in the wall.

For creeping flow, the first term on the right side in Eq. (2.139) is negligible. In this case, the pressure will decrease with increasing r. Assume that $a \ll r$. For creeping flow, Eq. (2.139) can approximately be written

$$\mu \frac{\partial^2 v_r}{\partial z^2} \approx \frac{\partial p}{\partial r} \tag{2.145}$$

The solution $v_r(r, z)$ can approximately be written

$$v_r(r, z) \approx \frac{3q}{8\pi ar} \left(1 - \frac{z^2}{a^2}\right) \tag{2.146}$$

$$\frac{\partial p}{\partial r} \approx -\frac{3\mu q}{4\pi a^3 r} \tag{2.147}$$

$$p \approx p_0 - \frac{3\mu q}{4\pi a^3} \ln \frac{2r}{D} \tag{2.148}$$

where p_0 is the pressure outside the disc in Fig. 2.5. The pressure difference Δp between $r = d/2$ and $r = D/2$ can approximately be written

$$\Delta p \approx \frac{3\mu q}{4\pi a^3} \ln \frac{D}{d} \tag{2.149}$$

If p_0 is the pressure outside the disc in Fig. 2.5, the pressure p_1 inside the disc between $r = 0$ and $r = d/2$ can approximately be written $p_1 \approx p_0 + \Delta p$. Then, the total force F on the disc in the z-axis direction can approximately be written

$$F \approx \frac{3\mu q}{32 a^3} \left(D^2 - d^2\right) \tag{2.150}$$

In this case, the force on the disc is directed out from the hole in the wall. It is very difficult to find a general and analytical solution to Eq. (2.139) for the problem above. The reason for this is that one must know the solution to be able to write the right side in Eq. (2.139) as a Fourier series in any orthogonal system.

One can do experiments with water or air where the disc is sucked towards the hole in the wall and stops at a certain distance from the wall. One can get the disc to float freely under the hole even when the gravitational field is directed in the z-axis direction.

2.7.6 Flow in an Annular Channel

Consider an annular channel containing a flowing incompressible fluid with constant viscosity μ where the channel is limited radially by two cylindrical surfaces with diameter d and D. Assume that the center line of the cylinders coincide with the z-axis in a cylindrical coordinate system as in Fig. 2.6. See also Fig. 2.2.

Assume that the flow is stationary and independent of the coordinates z and θ. This is valid for fully developed flow profile. Further, assume that all points which follow the flow, move with constant velocity along straight lines. The first term on the right side in Eq. (2.68) is, in this case, equal to zero since there are no acceleration forces in the flow. Then, in cylindrical coordinates, Eq. (2.68) can be written

$$\mu \frac{1}{r} \frac{\partial}{\partial r} \left(r \frac{\partial v_z}{\partial r} \right) = \frac{\partial p}{\partial z} \tag{2.151}$$

In this case, $\partial p / \partial z$ is a constant, independent of r and z. Then, Eq. (2.151) can be written

$$\frac{1}{r} \frac{\partial}{\partial r} \left(r \frac{\partial v_z}{\partial r} \right) = -C = -\frac{\Delta p}{\mu \, \Delta z} \tag{2.152}$$

where Δp is the pressure difference between two points with the distance Δz along the z-axis. C is a positive constant. Equation (2.152) is the Poisson equation with constant source term. Assume that $D = 2R$. Then, the solution $v_z(r)$ can be written

$$v_z = \frac{1}{4} \frac{\Delta p}{\mu \, \Delta z} R^2 \left[1 - (r/R)^2 + \frac{1 - (d/D)^2}{ln(D/d)} \ln (r/R) \right] \tag{2.153}$$

The volume flow q through the channel can be written

$$q = 2\pi \int_{d/2}^{D/2} v_z r \, dr = \frac{\pi}{8} \frac{\Delta p}{\mu \, \Delta z} R^4 \left[1 - (d/D)^4 - \frac{\left[1 - (d/D)^2 \right]^2}{\ln (D/d)} \right] \tag{2.154}$$

Fig. 2.6 Annular channel

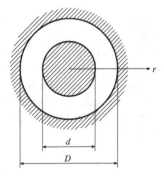

The hydraulic diameter D_h is defined

$$D_h = D - d \tag{2.155}$$

The linear velocity $\langle v \rangle$ is defined

$$\langle v \rangle = \frac{4q}{\pi \left(D^2 + d^2 \right)} \tag{2.156}$$

The Reynolds number Re is defined

$$Re = \frac{\rho \langle v \rangle D_h}{\mu} \tag{2.157}$$

The friction factor $f(Re)$ is defined

$$\Delta p = \frac{1}{2} \rho \langle v \rangle^2 \frac{4\Delta z}{D_h} f(Re) \tag{2.158}$$

Then, $f(Re)$ can be written

$$f(Re) = \frac{C_f}{Re} \tag{2.159}$$

$$C_f = \frac{16 \left(D^2 + d^2 \right) \left(D - d \right)^2 / D^4}{1 - (d/D)^4 - \dfrac{\left[1 - (d/D)^2 \right]^2}{\ln (D/d)}} \tag{2.160}$$

Note that the value of C_f is equal to 16 for a cylindrical channel with diameter D. Equation (2.159) is only valid for laminar flow.

2.7.7 Couette Flow Between Two Rotating Cylinders

Consider two concentrically arranged cylindrical surfaces with diameter D and d, where the cylinder center lines coincide with the z-axis in a cylindrical coordinate system as in Fig. 2.7. See also Fig. 2.2.

Assume that the space between the cylinders contains an incompressible fluid with constant viscosity μ and that the outer cylinder surface rotates with constant angular velocity ω. Further, assume that the cylinders are so long that the flow is independent of the z coordinate. The flow is also independent of θ. v_r and v_z are equal to zero. Then, in cylindrical coordinates, Eq. (2.68) can be written

$$0 = -\rho \frac{v_\theta^2}{r} + \frac{\partial p}{\partial r} \tag{2.161}$$

Fig. 2.7 Couette flow

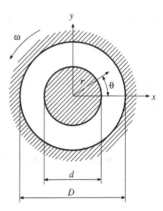

$$\frac{1}{r}\frac{\partial}{\partial r}\left(r\frac{\partial v_\theta}{\partial r}\right) - \frac{v_\theta}{r^2} = 0 \quad \text{or} \quad \frac{\partial}{\partial r}\left(\frac{\partial v_\theta}{\partial r} + \frac{v_\theta}{r}\right) = 0 \qquad (2.162)$$

Assume that $d = 2R$. Then, the solution $v_\theta(r)$ can be written

$$v_\theta(r) = \frac{\omega D^2}{D^2 - d^2}\left(r - \frac{R^2}{r}\right) \qquad (2.163)$$

If L is the length of the cylinders, the torque M between the cylinders can be written

$$M = \frac{1}{2}\pi\mu\omega\frac{d^2 D^2}{D^2 - d^2}L \qquad (2.164)$$

It is common that the viscosity of a liquid is determined from Eq. (2.164) by measuring the torque in a Couette viscometer. Equation (2.164) is only valid for laminar flow between the cylinders.

2.8 Two-Dimensional Potential Flow

For two-dimensional flow, the state variables only depend on two space coordinates. If these coordinates are x and y in a Cartesian coordinate system, the flow is a plane two-dimensional flow. If the flow is incompressible, the continuity equation Eq. (2.15) can be written

$$\frac{\partial v_x}{\partial x} + \frac{\partial v_y}{\partial y} = 0 \qquad (2.165)$$

The curl $\nabla \times \mathbf{v}$ of the velocity vector (v_x, v_y) can be written

$$\nabla \times \mathbf{v} = \hat{z}\left(\frac{\partial v_y}{\partial x} - \frac{\partial v_x}{\partial y}\right) \qquad (2.166)$$

The stream function Ψ is defined

$$v_x = \frac{\partial \Psi}{\partial y} \qquad v_y = -\frac{\partial \Psi}{\partial x} \tag{2.167}$$

The definition of the stream function Ψ means that the continuity equation Eq. (2.165) automatically is satisfied and that the curl $\nabla \times \mathbf{v}$ can be written

$$\nabla \times \mathbf{v} = -\hat{z}\left(\frac{\partial^2 \Psi}{\partial x^2} + \frac{\partial^2 \Psi}{\partial y^2}\right) \tag{2.168}$$

The definition of the stream function also means that the stream function Ψ is constant along a streamline. Streamlines with constant stream function are always orthogonal to curves with constant velocity potential. See Eq. (2.45). It is only possible to define a stream function for two-dimensional flow, but the velocity potential can be defined for three-dimensional flow. The volume flow q between the two streamlines 1 and 2 and two plane surfaces which are orthogonal to the z-axis can be written

$$q = \Delta z \left(\Psi_2 - \Psi_1\right) \tag{2.169}$$

where Δz is the distance between the plane surfaces and Ψ_1 and Ψ_2 are the values of the stream function at the both streamlines. Rotationless flow means that $\nabla \times \mathbf{v} = 0$. At stationary rotationless flow of an ideal incompressible fluid, $\nabla^2 \Phi = 0$ and $\nabla^2 \Psi = 0$, where Φ is the velocity potential. See Eqs. (2.51) and (2.168). The Laplace equation is valid for Φ and Ψ. This means that Φ and Ψ are harmonic functions. The function $\phi(z)$ is defined

$$\phi(z) = -\Phi(x, y) + i\Psi(x, y) \tag{2.170}$$

z in Eq. (2.170) must not be confused with the space coordinate z. Note that the minus sign in front of Φ in Eq. (2.170) is a consequence of that the velocity potential Φ is defined in such a way that $\mathbf{v} = -\nabla\Phi$ in this book. It happens that Φ is defined in such way that $\mathbf{v} = \nabla\Phi$. Therefore, it is common that Eq. (2.170) is written without any minus sign in front of Φ. It happens that the both definitions of the velocity potential are confused with each other in the same book. See also Eqs. (2.45) and (2.173). Since Φ and Ψ are harmonic functions, $\phi(z)$ is an analytic function of $z = x + iy$. The function $\phi(z)$ is called the complex velocity potential and represents the flow in the complex plane. If it is possible to find the analytic function $\phi(z)$, corresponding to the given boundary conditions, the solution of flow problem is known.

2.8.1 Potential Flow Around a Cylinder

Consider an infinitely long cylinder of diameter $2R$ which is located in a flow field of an incompressible ideal fluid as in Fig. 2.8. Assume that its center line coincides

Fig. 2.8 Potential flow
around a cylinder

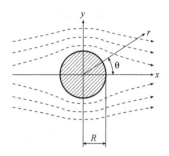

with the z-axis and the flow field is homogeneous with the velocity v_0 at infinite distance from the cylinder and is directed in the x-axis direction.

Assume that the complex velocity potential $\phi(z)$ in the complex plane $z = x + iy$ can be written

$$\phi(z) = v_0 \left(z + \frac{R^2}{z} \right) \tag{2.171}$$

z in Eq. (2.171) must not be confused with the space coordinate z. Division of $\phi(z)$ in a real part and an imaginary part gives

$$\phi(z) = v_0 \, x \, \left(1 + \frac{R^2}{x^2 + y^2} \right) + i \, v_0 \, y \, \left(1 - \frac{R^2}{x^2 + y^2} \right) =$$
$$v_0 \, r \, \cos\theta \left(1 + \frac{R^2}{r^2} \right) + i \, v_0 \, r \, \sin\theta \left(1 - \frac{R^2}{r^2} \right) \tag{2.172}$$

Equation (2.170) implies that the velocity potential Φ and the stream function Ψ can be written

$$- \Phi = v_0 \, x \, \left(1 + \frac{R^2}{x^2 + y^2} \right) = v_0 \, r \, \cos\theta \left(1 + \frac{R^2}{r^2} \right) \tag{2.173}$$

$$\Psi = v_0 \, y \, \left(1 - \frac{R^2}{x^2 + y^2} \right) = v_0 \, r \, \sin\theta \left(1 - \frac{R^2}{r^2} \right) \tag{2.174}$$

Note that the minus sign in front of Φ in Eq. (2.173) is a consequence of that the velocity potential Φ is defined in such way that $\mathbf{v} = -\nabla\Phi$ in this book. It happens that Φ is defined in such way that $\mathbf{v} = \nabla\Phi$. See also Eqs. (2.45) and (2.170). Equations (2.45) and (2.167) means that v_x and v_y can be written

$$v_x = -\frac{\partial\Phi}{\partial x} = \frac{\partial\Psi}{\partial y} = v_0 \left(1 - \frac{R^2}{r^2} \cos 2\theta \right) \tag{2.175}$$

$$v_y = -\frac{\partial\Phi}{\partial y} = -\frac{\partial\Psi}{\partial x} = - v_0 \frac{R^2}{r^2} \sin 2\theta \tag{2.176}$$

Equations (2.175) and (2.176) means that v_x approaches v_0 and v_y approaches 0 when r approaches infinity. Equation (2.174) means that $\Psi = 0$ on the cylinder surface. Since Ψ is constant at a streamline, this means that $\phi(z)$ satisfies the boundary conditions for the flow problem and Eqs. (2.175) and (2.176) is the solution to the flow problem. The polar components v_r and v_θ of the velocity vector can be written

$$v_r = v_0 \left(1 - \frac{R^2}{r^2}\right) \cos\theta \qquad (2.177)$$

$$-v_\theta = v_0 \left(1 + \frac{R^2}{r^2}\right) \sin\theta \qquad (2.178)$$

2.9 Boundary Layer

For frictionless flow around a solid object, the velocity vector normal component is zero at the surface of the object. However, there is no boundary condition for the tangential component of the velocity vector. In a real fluid, where the viscosity μ is different from zero, the tangential component of the velocity vector is also zero at the surface of the object. In many cases, the viscosity influence on the flow is concentrated in the *boundary layer* next to the surface of the object. Outside the boundary layer, the flow can approximately be regarded as potential flow. This is not valid for the streamlines which go from the boundary layer and downstream form the wake of the object. In a laminar boundary layer, the friction forces are dominant, while the acceleration forces dominate the flow outside the boundary layer. The transition between the boundary layer flow and the potential flow is continuous. We must therefore define the boundary layer thickness with any convention. In practice, it is expected that the velocity will be 99 % of the velocity outside the boundary layer at the interface between the boundary layer and the flow outside the boundary layer. Consider an infinitely large plate which coincides with the x-z-plane as in Fig. 2.9. Assume that there is a tangential flow of an incompressible fluid with constant viscosity μ where the velocity field $\mathbf{v}$ is homogeneous and directed in the x-axis direction at infinite distance from the plate. Further, assume that the absolute value of $\mathbf{v}$ is equal to v_0 at infinite distance from the plate and that the gravitational field $g_i = 0$. The following properties are valid for the boundary layer.

$$v_x \gg v_y \qquad (2.179)$$

Fig. 2.9 Laminar boundary layer

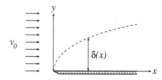

$$\frac{\partial v_x}{\partial y} \gg \frac{\partial v_x}{\partial x} = -\frac{\partial v_y}{\partial y} \tag{2.180}$$

One can show that the pressure p is approximately independent of y in the boundary layer. $p \approx p_0(x)$ where $p_0(x)$ is the pressure in the external flow outside the boundary layer. Under these conditions, Eq. (2.68) is reduced to the boundary layer equation.

$$v_x \frac{\partial v_x}{\partial x} + v_y \frac{\partial v_x}{\partial y} = -\frac{1}{\rho}\frac{dp_0}{dx} + \nu \frac{\partial^2 v_x}{\partial y^2} \quad \nu = \mu/\rho \tag{2.181}$$

The continuity equation Eq. (2.15) can be written

$$\frac{\partial v_x}{\partial x} + \frac{\partial v_y}{\partial y} = 0 \tag{2.182}$$

The boundary conditions can be written

$$v_x(x, 0) = v_y(x, 0) = 0 \tag{2.183}$$

$$\lim_{y \to \infty} v_x(x, y) = v_0 \tag{2.184}$$

v_x is approximately constant, independent of x and equal to v_0 outside the boundary layer. Then, the Bernoulli equation Eq. (2.33) leads to $dp_0/dx = 0$. Then, Eq. (2.181) can be written

$$v_x \frac{\partial v_x}{\partial x} + v_y \frac{\partial v_x}{\partial y} = \nu \frac{\partial^2 v_x}{\partial y^2} \tag{2.185}$$

The definition of the stream function Ψ (2.167) implies that the continuity equation Eq. (2.182) automatically is satisfied. Equation (2.167) means that Eq. (2.185) can be written

$$\frac{\partial \Psi}{\partial y}\frac{\partial^2 \Psi}{\partial x \partial y} - \frac{\partial \Psi}{\partial x}\frac{\partial^2 \Psi}{\partial y^2} = \nu \frac{\partial^3 \Psi}{\partial y^3} \tag{2.186}$$

The boundary conditions can be written

$$\frac{\partial \Psi}{\partial x} = \frac{\partial \Psi}{\partial y} = 0 \quad \text{when } y = 0 \tag{2.187}$$

$$\lim_{y \to \infty} \frac{\partial \Psi}{\partial y} = v_0 \tag{2.188}$$

Assume that $\Psi(x, y)$ can be written

$$\Psi(x, y) = \left(\nu v_0 x\right)^{1/2} f(\eta) \tag{2.189}$$

$$f(\eta) = \frac{v_x(\eta)}{v_0} \tag{2.190}$$

$$\eta = \frac{y}{\delta(x)} \tag{2.191}$$

Then, Eq. (2.186) can be written

$$\frac{1}{\delta(x)} \left(\frac{v\,x}{v_0} \right)^{1/2} f'''(\eta) + \frac{1}{2} f(\eta)\, f''(\eta)$$
$$= \left(\frac{1}{2} - \frac{x}{\delta(x)} \frac{\partial\,\delta(x)}{\partial x} \right) [f'(\eta)]^2 \tag{2.192}$$

If $f(\eta)$ only depends on η, the parenthesis on the right side in Eq. (2.192) and the factor in front of $f'''(\eta)$ in the same equation must be independent of x. This condition is satisfied if the parenthesis on the right side is zero, which means that the following equation is valid.

$$\frac{\partial\,\delta(x)}{\partial x} = \frac{\delta(x)}{2\,x} \tag{2.193}$$

The solution to Eq. (2.193) can be written

$$\delta(x) = C\,x^{1/2} \tag{2.194}$$

where C is a constant, independent of x and η. If the solution $\Psi(x, y)$ can be written as Eq. (2.189), Eq. (2.194) must be valid. Then, Eq. (2.192) can be written

$$f'''(\eta) + \frac{1}{2} C\,(v_0/v)^{1/2}\, f(\eta)\, f''(\eta) = 0 \tag{2.195}$$

If $C = (v/v_0)^{1/2}$, the following equations are valid.

$$f'''(\eta) + \frac{1}{2} f(\eta)\, f''(\eta) = 0 \tag{2.196}$$

$$\delta(x) = \left(\frac{v\,x}{v_0} \right)^{1/2} \tag{2.197}$$

Equation (2.196) is called the *Blasius equation*. The boundary conditions can be written

$$f(0) = f'(0) = 0 \tag{2.198}$$

$$\lim_{\eta \to \infty} f'(\eta) = 1 \tag{2.199}$$

It is not trivial to solve Eq. (2.196). The equation can only be solved numerically.

2.10 Turbulent Flow

The general idea of turbulent flow is that laminar flow for high Reynolds number becomes unstable and turns into a chaotic process in which the velocity vector and the pressure varies stochastically in time. For turbulent flow, the acceleration forces dominate over the friction forces in the flow. This means that the first term on the right side in Eq. (2.17) is much greater than the last term on the left side in i Eq. (2.17). Consider turbulent flow in an incompressible fluid with constant viscosity μ. Assume that $\bar{v}_i$ is the expectation value of v_i and $\bar{p}$ is the expectation value of p. Then, v_i and p can be written

$$v_i = \bar{v}_i + v_i' \tag{2.200}$$
$$p = \bar{p} + p' \tag{2.201}$$

where v_i' and p' are time dependent stochastic variables. Note that $\bar{v}_i$ and $\bar{p}$ are independent of the time at stationary turbulent flow. For non-stationary turbulent flow, the expectation value is an average of the value at the same time for different realizations of the stochastic process when the number of realizations approaches infinity. This is also valid for stationary turbulent flow, but in this case, the expectation value is also equal to the time average when the time approaches infinity. For incompressible flow, the continuity equation Eq. (2.15) can be written

$$\left(\bar{v}_i + v_i'\right)_{,i} = 0 \tag{2.202}$$

Equation (2.19) can be written

$$\mu \left(\bar{v}_{i,jj} + v_{i,jj}'\right)$$
$$= \rho \left(\frac{\partial \bar{v}_i}{\partial t} + \bar{v}_j \bar{v}_{i,j} - g_i\right) + \bar{p}_{,i}$$
$$+ \rho \left(\frac{\partial v_i'}{\partial t} + v_j' v_{i,j}' + \bar{v}_j v_{i,j}' + v_j' \bar{v}_{i,j}\right) + p_{,i}' \tag{2.203}$$

The expectation value of v_i' and p' is by definition equal to zero and the expectation value of $\bar{v}_i$ and $\bar{p}$ is equal to $\bar{v}_i$ and $\bar{p}$. Then, the expectation value of Eq. (2.202) gives the following equation.

$$\bar{v}_{i,i} = 0 \tag{2.204}$$

Equations (2.202) and (2.204) give the following equation.

$$v_{i,i}' = 0 \tag{2.205}$$

Equation (2.205) means that $v'_j\, v'_{i,j}$ in Eq. (2.203) can be written

$$v'_j\, v'_{i,j} = \left(v'_i\, v'_j\right)_{,j} \tag{2.206}$$

Then, the expectation value of Eq. (2.203) gives the following equation.

$$\mu\, \overline{v}_{i,jj} = \rho \left(\frac{\partial \overline{v}_i}{\partial t} + \overline{v}_j\, \overline{v}_{i,j} + \underbrace{\overline{\left(v'_i\, v'_j\right)}_{,j}}_{\text{turbulence}} - g_i \right) + \overline{P}_{,i} \tag{2.207}$$

Equation (2.207) is identical to Eq. (2.19) except on the third term in the parenthesis in Eq. (2.207). For stationary turbulent flow, $\overline{\left(v'_i\, v'_j\right)}_{,j}$ is a time independent function which depends on the space coordinates $\mathbf{x}$. One can not predict how this function looks if one does not know the time dependent solution to Eq. (2.19) for stationary turbulent flow. Experiments show that v_i and p usually vary in time at stationary turbulent flow. In some experiments where the flow is very stable, the flow will be a "turbulent" flow with stable time independent vortices. However, it is doubtful to consider such a state as turbulent flow. It is not entirely clear how to define turbulent flow. Consider stationary flow in a cylindrical tube with fully developed flow profile. It is well known that the flow changes from laminar flow to turbulent flow in smooth pipes when the Reynolds number $Re \approx 2100$. At highly controlled experimental conditions, the flow can be laminar even when Re is greater than 2100. One talks about transition flow where there are elements of both laminar and turbulent flow. At transition flow, the value of Re is between 2100 and 4000. When Re is greater than 4000, one talks about fully developed turbulent flow. Assume that all points in the fluid move with constant velocity along straight lines in a cylindrical tube. Then, the first term on the right side in Eq. (2.68) is equal to zero since there are no acceleration forces in the flow. Then, the flow profile can be described by Eq. (2.93). The flow profile has the shape of a parabola. This solution is valid for all values of the Reynolds number even if the solution is unstable for high values of the Reynolds number. It is well known that the flow profile of the expectation value of the axial velocity for turbulent flow in a tube is more flattened than in laminar flow. It is very difficult to find analytical solutions to the Navier-Stokes equation if one can not neglect the friction forces or the acceleration forces even for very simple geometries. Therefore, it is impossible to find analytical solutions which describe turbulent flow. Turbulent flow is very much about the interplay between acceleration forces and friction forces. The only way to get a handle on turbulent flow seems to be to solve the Navier-Stokes equation numerically. Since the problem is very complex, it requires very powerful computers to do this. The computer development today has come so far that this is fully realistic. If one does not know the results of such calculations, one can only speculate about the nature of turbulent flow. At least one of the following statements is true.

- If the Reynolds number is less than a certain value, it only exists one time independent solution. This solution describes laminar flow. If the Reynolds number is greater than a certain value, it does not exist any stable time independent solution. Then, the flow must vary with time in some way. This variation can be periodic or chaotic.

- If the Reynolds number is less than a certain value, it only exists one time independent solution. This solution describes laminar flow. If the Reynolds number is greater than a certain value, it exists several time independent solutions where the original solution is unstable. These solutions are very sensitive to disturbances, which can give rise to a stochastic time dependent flow.

It is not obvious that the flow must be free from acceleration at stationary time independent flow with fully developed flow profile in a cylindrical tube. Note that the inflow process give rise to acceleration forces even at time independent flow. There may be solutions where the acceleration term in the Navier-Stokes equation is not zero even if the flow is time independent for fully developed flow profile. Then, it must exist stable time independent vortices in the tube. One realizes that these solutions are very complicated. It is possible that these solutions do not exist below a certain value of the Reynolds number Re. If it can exist stable time independent vortices in the tube, it is random where they will be located along the z-axis. One can imagine that small disturbances in the flow can have the vortices to propagate along the z-axis. Then, the flow becomes quasi stationary. The flow becomes more chaotic with increasing disturbances in the flow. The flow also becomes more chaotic with increasing Reynolds number. It is also possible that it does not exist stable time independent solutions over a certain value of Re. Then, the flow must vary with time in some way, even when the system is undisturbed. The time dependence can either be periodic or chaotic. It is only numerical experiments which can clarify how it works. There have been many experiments which form the basis of various correlations for turbulent flow. Such an example is the Blasius formula (2.103). For a very long time, it has only been possible to obtain information about turbulent flow through experiments.

2.11 Flow Separation

Consider an infinitely large plate which coincides with the x-z-plane as in Fig. 2.10. Assume that there is a tangential flow of an incompressible fluid with constant viscosity μ where the velocity field $\mathbf{v}$ is homogeneous and directed in the x-axis direction at infinite distance from the plate. Further, assume that the absolute value of $\mathbf{v}$ is equal to v_0 at infinite distance from the plate and that the gravitational field is equal to zero. In this case, the flow turns into turbulent flow at some value of the coordinate x.

Assume that the fluid is incompressible with constant viscosity μ and that the Reynolds number Re_x is defined

Fig. 2.10 Laminar-turbulent
transition on a plate

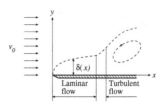

$$Re_x = \frac{v_0\,x}{v} \quad v = \mu/\rho \tag{2.208}$$

The transition to turbulent flow occurs when $Re_x \approx 5 \cdot 10^5$. In the laminar region, the boundary layer Eq. (2.185) is approximately valid. In the turbulent region, v_x and v_y can be written

$$v_x = \bar{v}_x + v'_x \tag{2.209}$$
$$v_y = \bar{v}_y + v'_y \tag{2.210}$$

where $\bar{v}_x$ and $\bar{v}_y$ are the expectation values of v_x and v_y. v'_x and v'_y are time dependent stochastic variables. Equations (2.204) and (2.205) give the following equations.

$$\frac{\partial \bar{v}_x}{\partial x} + \frac{\partial \bar{v}_y}{\partial y} = 0 \tag{2.211}$$

$$\frac{\partial v'_x}{\partial x} + \frac{\partial v'_y}{\partial y} = 0 \tag{2.212}$$

Equation (2.211) is the continuity equation. Then, the boundary layer equation in the turbulent region can be written

$$\bar{v}_x \frac{\partial \bar{v}_x}{\partial x} + \bar{v}_y \frac{\partial \bar{v}_x}{\partial y} + \underbrace{\frac{\partial}{\partial y}\,\overline{(v'_x\,v'_y)} + \frac{\partial}{\partial x}\,\overline{(v'_x)^2}}_{\text{turbulence}} = v\,\frac{\partial^2 \bar{v}_x}{\partial y^2} \tag{2.213}$$

Note that Eq. (2.213) assumes that Eq. (2.212) is valid. Two terms on the left side in the boundary layer equation are added for turbulent flow. These are caused by the turbulent motions and represent derivatives of the turbulent stresses. In a turbulent boundary layer, the first of these terms has a dominant influence on the flow. It is therefore common that the boundary layer equation for turbulent flow is written

$$\bar{v}_x \frac{\partial \bar{v}_x}{\partial x} + \bar{v}_y \frac{\partial \bar{v}_x}{\partial y} + \frac{\partial}{\partial y}\,\overline{(v'_x\,v'_y)} = v\,\frac{\partial^2 \bar{v}_x}{\partial y^2} \tag{2.214}$$

The turbulent shear stress τ_{turb} is defined

$$\tau_{turb} = -\rho\,\overline{(v'_x\,v'_y)} \tag{2.215}$$

Fig. 2.11 Flow separation
behind a cylinder

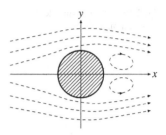

τ_{turb} is valid in the x-z-plane. Figure 2.8 shows potential flow around a cylinder. For a
real fluid with the viscosity μ which is different from zero, the solution for potential
flow is often fairly well valid at the front of the cylinder while there is turbulence
at the rear of the cylinder. There is a transition from laminar to turbulent boundary
layer at some point on the cylinder surface. The turbulent boundary layer after this
point is substantially thicker than the turbulent boundary layer on a plate. One talks
about formation of a wake behind the cylinder. Figure 2.11 shows schematically how
this may look. This applies to all kinds of geometric objects.

2.12 Flow in a Packed Bed

The total force on the particles in a control volume in a packed bed is caused by
friction and pressure forces on the particle surface. If the control volume V in Eq. (2.1)
contains fixed particles with the outer boundary surface S_p, the equation is replaced
by the following equation.

$$\frac{\partial}{\partial t} \int_V \rho\, v_i\, dV - \int_V \rho\, g_i\, dV$$

$$= - \int_S \rho\, v_i\, v_j\, n_j\, dS + \int_S \sigma_{ji}\, n_j\, dS - \int_{S_p} \sigma_{ji}\, n_j\, dS \qquad (2.216)$$

The unit normal n_i is directed out from the closed surfaces S and S_p. Assume that
the flow is stationary and that the momentum flow and the friction stress τ_{ij} are
negligible on the surface S. Then, Eq. (2.216) can be written

$$0 = \int_S p\, \delta_{ij}\, n_j\, dS - \int_{S_p} \sigma_{ji}\, n_j\, dS \qquad (2.217)$$

The Gauss theorem (1.31) and Eq. (2.10) means that the following equation is valid.

$$\int_S p \, \delta_{ij} \, n_j \, dS = \int_V p_{,i} \, dV \qquad (2.218)$$

Then, the total force F_i on the particles in the volume V can be written

$$F_i = -\int_{S_p} \sigma_{ji} \, n_j \, dS = -\int_V p_{,i} \, dV \qquad (2.219)$$

Assume that dS_p is the outer surface of the particles in an infinitesimal volume element dV in the volume V. This makes sense if the pressure can be regarded as constant in a volume element where the size of the volume element is of the same order of magnitude as the particle size. Further, assume that $dF_i = F_i \, dV$ where dF_i is the force on the particles in the volume dV. Then, Eq. (2.219) can be written

$$p_{,i} = -\frac{dF_i}{dV} \qquad (2.220)$$

To calculate the pressure gradient $p_{,i}$, it is necessary to know the flow pattern between the particles in the bed. It is impossible to find an analytical solution to the flow equation if the bed does not consist of straight channels which are directed in the flow direction. It is also very difficult to calculate the flow pattern in a packed bed with numerical methods. In a packed bed, both the friction and acceleration forces have great influence on the flow pattern even at quite low values of the Reynolds number. Typical values of the Reynolds number in a packed bed is between 50 and 100. Since the fluid is subjected to acceleration forces when it flows in curved paths between the particles, the flow profile between the particles will be highly flow dependent. Assume that the bed consists of straight cylindrical channels with diameter D which are oriented in the flow direction. Further, assume that ε_b is the porosity of the bed. Then, the Hagen-Poiseuille law (2.98) means that $p_{,i}$ can be written

$$p_{,i} = -\widehat{x}_i \frac{32 \, \mu \, G}{\varepsilon_b \, \rho \, D^2} \qquad (2.221)$$

if the channels are oriented in the x_i coordinate direction. G is the total mass flow divided by the cross-sectional area of the bed. In this case, the specific area S_v of the bed can be written

$$S_v = \frac{4 \, \varepsilon_b}{D} \qquad (2.222)$$

Assume that the bed consists of spherical particles with diameter D_p. Then, the specific area S_v of the bed can be written

$$S_v = \frac{6 \, (1 - \varepsilon_b)}{D_p} \qquad (2.223)$$

Assume that the specific areas S_v in Eqs. (2.222) and (2.223) are equal. Then, the following relation between D and D_p is valid.

$$D = \frac{2\,\varepsilon_b}{3\,(1 - \varepsilon_b)}\,D_p \tag{2.224}$$

Insertion of Eq. (2.224) into Eq. (2.221) gives

$$p_{,i} = -\widehat{x}_i\,\frac{72\,\mu\,G}{\rho\,D_p^2}\,\frac{(1 - \varepsilon_b)^2}{\varepsilon_b^3} \tag{2.225}$$

Equation (2.225) gives too low value of the pressure gradient for a packed bed of spheres. Experiments at low flow rates have shown that the right side in Eq. (2.225) shall be multiplied by the factor $25/12$. Then, the pressure gradient can be written

$$p_{,i} = -\widehat{x}_i\,\frac{150\,\mu\,G}{\rho\,D_p^2}\,\frac{(1 - \varepsilon_b)^2}{\varepsilon_b^3} \tag{2.226}$$

Equation (2.226) is called the *Blake-Kozeny equation*. The equation is approximately valid if the following condition is satisfied.

$$\frac{D_p\,G}{\mu}\,\frac{1}{(1 - \varepsilon_b)} < 10 \tag{2.227}$$

The friction factor $f(Re)$ for a packed bed is defined

$$p_{,i} = -\widehat{x}_i\,\frac{1}{2}\,\rho\,v_0^2\,\frac{4\,f(Re)}{D_p} \qquad v_0 = G/\rho \tag{2.228}$$

The Blake-Kozeny equation means that the friction factor $f(Re)$ can be written

$$f(Re) = \frac{(1 - \varepsilon_b)^2}{\varepsilon_b^3}\,\frac{75}{Re} \tag{2.229}$$

$$Re = \frac{D_p\,G}{\mu} \tag{2.230}$$

Already when $Re/(1 - \varepsilon_b) = 50$, Eq. (2.226) differs strongly from observed values. The reason for this is that Eq. (2.226) does not take into account the acceleration forces in the flow. It is very rare that the flow is turbulent in a packed bed. At highly turbulent flow, the following equation is approximately valid.

$$p_{,i} = -\widehat{x}_i\,3.50\,\frac{1}{2}\,\rho\,v_0^2\,\frac{1}{D_p}\,\frac{1 - \varepsilon_b}{\varepsilon_b^3} \qquad v_0 = G/\rho \tag{2.231}$$

$$f(Re) = 0.875 \frac{1 - \varepsilon_b}{\varepsilon_b^3} \qquad \frac{Re}{1 - \varepsilon_b} > 1000 \qquad (2.232)$$

Equation (2.231) is called the *Burke-Plummer equation*. Addition of the Blake-Kozeny equation and the Burke-Plummer equation gives the following equation.

$$p_{,i} = -\widehat{x}_i \frac{1 - \varepsilon_b}{\rho\, D_p\, \epsilon_b^3} \left[150 \frac{\mu\, (1 - \varepsilon_b)}{D_p} + 1.75\, G \right] G \qquad (2.233)$$

Equation (2.233) is called the *Ergun equation*. The Ergun equation is approximately valid for homogeneous flow in a packed bed of spherical particles. The equation gives larger errors if $Re/(1 - \varepsilon_b) > 10$ than if $Re/(1 - \varepsilon_b) < 10$. The reason for this is that the flow is not turbulent in the whole region where $Re/(1 - \varepsilon_b) > 10$. It is not entirely correct to add the Blake-Kozeny equation and the Burke-Plummer equation.

2.13 Acoustic Impedance of Audio Sources

This section describes the acoustic impedance of some common audio sources. Assume that the value of the velocity potential Φ in Eq. (2.62) is equal to 0 when $t \leqslant 0$. Then, Laplace transformation of Eqs. (2.62), (2.63) and (2.64) gives the following equations.

$$\nabla^2 \widetilde{\Phi} = \frac{s^2}{c^2} \widetilde{\Phi} \qquad (2.234)$$

$$\widetilde{\mathbf{v}} = -\nabla \widetilde{\Phi} \qquad (2.235)$$

$$\widetilde{\Delta p} = \rho_0 s\, \widetilde{\Phi} \qquad (2.236)$$

Consider a surface S which moves with the velocity $\mathbf{v} = v\,\widehat{\mathbf{n}}$ where $\widehat{\mathbf{n}}$ is the unit normal on the surface S. The volume flow through the surface can be written

$$q = \int_S v\,\widehat{\mathbf{n}} \cdot d\mathbf{S} \qquad (2.237)$$

The movement gives rise to the sound pressure Δp on the surface S. The total force F on the surface is defined

$$F = \int_S \Delta p\,\widehat{\mathbf{n}} \cdot d\mathbf{S} \qquad (2.238)$$

Laplace transformation of (2.237) and (2.238) gives the following equations.

$$\widetilde{q} = \int_S \widetilde{v}\,\widehat{\mathbf{n}} \cdot d\mathbf{S} \tag{2.239}$$

$$\widetilde{F} = \int_S \widetilde{\Delta p}\,\widehat{\mathbf{n}} \cdot d\mathbf{S} \tag{2.240}$$

The mechanical impedance Z_M of the surface S is defined

$$\widetilde{F} = Z_M \widetilde{q}/A \tag{2.241}$$

where A is the area of S. The acoustic impedance Z_A of the surface S is defined

$$\widetilde{F} = Z_A \widetilde{q} A \tag{2.242}$$

The mechanical impedance Z_M is expressed as a force divided by a velocity. The acoustic impedance Z_A is expressed as a pressure divided by a volume flow.

2.13.1 Plane Wave Radiation

Consider an infinitely long straight channel with constant cross-sectional area which is oriented in the x coordinate direction. Assume that there is a rigid oscillating surface with the area A at $x = 0$ which is oriented perpendicular to the x-axis. Then, the Laplace transform $\widetilde{\Phi}$ of the velocity potential will be a function of x and s. Then, Eq. (2.234) can be written

$$\frac{\partial^2}{\partial x^2}\widetilde{\Phi} = \frac{s^2}{c^2}\widetilde{\Phi} \tag{2.243}$$

The solution to Eq. (2.243) can be written

$$\widetilde{\Phi} = C_1 e^{-sx/c} + C_2 e^{sx/c} \tag{2.244}$$

The first term on the right side in Eq. (2.244) describes a wave propagation in the x coordinate direction and the second term describes a wave propagation in the opposite direction. This means that $C_2 = 0$ in this case. Then, the Laplace transform $\widetilde{\mathbf{v}}$ of the velocity $\mathbf{v}$ can be written

$$\widetilde{\mathbf{v}} = -\frac{\partial \widetilde{\Phi}}{\partial x} = \widehat{\mathbf{x}}\,C_1\frac{s}{c}e^{-sx/c} \tag{2.245}$$

The Laplace transform $\widetilde{\Delta p}$ of the sound pressure Δp can be written

$$\widetilde{\Delta p} = \rho_0 s\,\widetilde{\Phi} = \rho_0 s\,C_1 e^{-sx/c} \tag{2.246}$$

The Laplace transform $\widetilde{q}$ of the volume flow q at $x = 0$ can be written

$$\widetilde{q} = A\,C_1 \frac{s}{c} \tag{2.247}$$

The Laplace transform $\widetilde{F}$ of the force F at $x = 0$ can be written

$$\widetilde{F} = A\,C_1 \rho_0 s \tag{2.248}$$

Then, the impedances Z_M and Z_A can be written

$$Z_M = A\,\rho_0\,c \tag{2.249}$$
$$Z_A = \rho_0\,c/A \tag{2.250}$$

Note that the imaginary part of Z_M and Z_A in this case is equal to zero, while the real part is a constant independent of the frequency. The real part is called the *radiation resistance*.

2.13.2 Pulsating Sphere

Consider a pulsating spherical surface S with the area A and the radius r_0 which is located in an infinite space. Assume that $\mathbf{v} = v\,\widehat{\mathbf{r}}$ is the velocity of the sphere where v is independent of the coordinates θ and φ in spherical coordinates. Then, the Laplace transform $\widetilde{\Phi}$ of the velocity potential Φ will be a function of r and s. Then, Eq. (2.234) can be written

$$\frac{\partial^2}{\partial r^2}\left(r\,\widetilde{\Phi}\right) = \frac{s^2}{c^2}\left(r\,\widetilde{\Phi}\right) \tag{2.251}$$

The solution to Eq. (2.251) can be written

$$\widetilde{\Phi} = \frac{C_1}{r}\,e^{-sr/c} + \frac{C_2}{r}\,e^{sr/c} \tag{2.252}$$

The first term on the right side in Eq. (2.252) describes a wave propagation in the r coordinate direction and the second term describes a wave propagation in the opposite direction. This means that $C_2 = 0$ in this case. Then, the Laplace transform $\widetilde{\mathbf{v}}$ of the velocity $\mathbf{v}$ can be written

$$\widetilde{\mathbf{v}} = -\frac{\partial\widetilde{\Phi}}{\partial r} = \widehat{\mathbf{r}}\frac{C_1}{r^2}\left(1 + \frac{sr}{c}\right)e^{-sr/c} \tag{2.253}$$

The Laplace transform $\widetilde{\Delta p}$ of the sound pressure Δp can be written

$$\widetilde{\Delta p} = \rho_0 s \, \widetilde{\Phi} = C_1 \, \rho_0 \frac{s}{r} e^{-s r/c} \tag{2.254}$$

The Laplace transform $\widetilde{q}$ of the volume flow q at $r = r_0$ can be written

$$\widetilde{q} = A \frac{C_1}{r_0^2} \left(1 + \frac{s \, r_0}{c} \right) e^{-s r_0/c} \tag{2.255}$$

The Laplace transform $\widetilde{F}$ of the force F at $r = r_0$ can be written

$$\widetilde{F} = A \, C_1 \, \rho_0 \frac{s}{r_0} e^{-s r_0/c} \tag{2.256}$$

Then, the impedances Z_M and Z_A can be written

$$Z_M = A \, \rho_0 c \cdot \frac{s r_0/c}{1 + s r_0/c} \tag{2.257}$$

$$Z_A = \frac{\rho_0 c}{A} \cdot \frac{s r_0/c}{1 + s r_0/c} \tag{2.258}$$

Z_M and Z_A have such properties that the integrand in the Bromwich-Wagner integral only has singular points to the left of the imaginary axis. Then, Z_M can be written

$$Z_M = A \, \rho_0 \, c \cdot \frac{i \, \omega \, r_0/c}{1 + i \, \omega \, r_0/c} = A \, \rho_0 \, c \cdot \frac{(\omega \, r_0/c)^2 + i \, \omega \, r_0/c}{1 + (\omega \, r_0/c)^2} \tag{2.259}$$

Z_M can also be written

$$Z_M = R_S + i \, \omega \, m_S \tag{2.260}$$

$$R_S = A \, \rho_0 \, c \cdot \frac{(\omega \, r_0/c)^2}{1 + (\omega \, r_0/c)^2} \tag{2.261}$$

$$m_S = \frac{A \, \rho_0 \, r_0}{1 + (\omega r_0/c)^2} \tag{2.262}$$

where R_S is the radiation resistance and m_S is the medium moving mass.

2.13.3 Oscillating Piston in an Infinite Baffle

Consider a rigid oscillating circular piston with the radius r_0 and the area A which is placed in an infinite wall or baffle. The mechanical impedance of such an audio source can be written

$$Z_M = A \rho_0 c \left[1 - \frac{J_1(2\omega r_0/c)}{\omega r_0/c} \right] + i \frac{\pi \rho_0 c^3}{2\omega^2} K_1(2\omega r_0/c) \qquad (2.263)$$

where J_1 and K_1 are Bessel functions of the first and second kind.

$$1 - \frac{J_1(2\omega r_0/c)}{\omega r_0/c}$$

$$= \frac{(\omega r_0/c)^2}{2} - \frac{(\omega r_0/c)^4}{2^2 3} + \frac{(\omega r_0/c)^6}{2^2 3^2 4} - \cdots \qquad (2.264)$$

$$K_1(2\omega r_0/c)$$

$$= \frac{2}{\pi} \left[\frac{(2\omega r_0/c)^3}{3} - \frac{(2\omega r_0/c)^5}{3^2 5} + \frac{(2\omega r_0/c)^7}{3^2 5^2 7} - \cdots \right] \qquad (2.265)$$

Z_M can also be written

$$Z_M = R_s + i\omega m_s \qquad (2.266)$$

$$R_s = A \rho_0 c \left[1 - \frac{J_1(2\omega r_0/c)}{\omega r_0/c} \right] \qquad (2.267)$$

$$m_s = \frac{\pi \rho_0 c^3}{2\omega^3} K_1(2\omega r_0/c) \qquad (2.268)$$

where R_s is the radiation resistance and m_s is the medium moving mass. For small values of $\omega r_0/c$, R_s and m_s can be written

$$R_s = \frac{A^2 \rho_0 \omega^2}{2\pi c} \qquad (2.269)$$

$$m_s = \frac{8}{3} \rho_0 r_0^3 \qquad (2.270)$$

Equations (2.269) and (2.270) are valid when the wavelength is much greater than the diameter of the piston. Note that R_s is proportional to the square of the area in this case.

2.13.4 Exponential Horn

Consider an infinitely long slightly diverging circular channel where the channel center line coincides with the x-axis. Assume that the channel cross-sectional area A varies with x as

$$A = A_0 e^{ax} \qquad (2.271)$$

where a is a constant and A_0 is the channel cross-sectional area at the throat ($x = 0$). For axial wave propagation, the following equation is approximately valid.

$$\frac{\partial^2 \Phi}{\partial x^2} + a \frac{\partial \Phi}{\partial x} = \frac{1}{c^2} \frac{\partial^2 \Phi}{\partial t^2} \tag{2.272}$$

Φ is the velocity potential. See Eq. (2.45). Laplace transformation of Eq. (2.272) gives

$$\frac{\partial^2 \tilde{\Phi}}{\partial x^2} + a \frac{\partial \tilde{\Phi}}{\partial x} = \frac{s^2}{c^2} \tilde{\Phi} \tag{2.273}$$

where $\tilde{\Phi}$ is the Laplace transform of Φ. The solution to Eq. (2.273) can be written

$$\tilde{\Phi} = e^{-ax/2} \left(C_1 e^{-\gamma s x/c} + C_2 e^{\gamma s x/c} \right) \tag{2.274}$$

$$\gamma = \sqrt{1 + \frac{a^2 c^2}{4 s^2}} \tag{2.275}$$

The first term in the parenthesis in Eq. (2.274) describes a wave propagation in the x coordinate direction and the second term describes a wave propagation in the opposite direction. This means that $C_2 = 0$ in this case. Then, the Laplace transform $\tilde{\mathbf{v}}$ of the velocity $\mathbf{v}$ can be written.

$$\tilde{\mathbf{v}} = -\frac{\partial \tilde{\Phi}}{\partial x} = \hat{\mathbf{x}} C_1 \left(\frac{\gamma s}{c} + \frac{a}{2} \right) e^{-ax/2} e^{-\gamma s x/c} \tag{2.276}$$

The Laplace transform $\widetilde{\Delta p}$ of the sound pressure Δp can be written

$$\widetilde{\Delta p} = \rho_0 s \tilde{\Phi} = \rho_0 s C_1 e^{-ax/2} e^{-\gamma s x/c} \tag{2.277}$$

Then, the impedance Z_M at the throat of the exponential horn can be written

$$Z_M = \frac{A_0 \widetilde{\Delta p}}{\tilde{v}_x} = A_0 \rho_0 c \left(\gamma - \frac{a c}{2 s} \right) \tag{2.278}$$

Z_M has such properties that the integrand in the Bromwich-Wagner integral only has singular points to the left of the imaginary axis. Then, Z_M can be written

$$Z_M = A_0 \rho_0 c \left(\sqrt{1 - \frac{a^2 c^2}{4 \omega^2}} + i \frac{a c}{2 \omega} \right) \tag{2.279}$$

Z_M can also be written

$$Z_M = R_s + i \omega m_s \tag{2.280}$$

$$R_s = A_0 \rho_0 c \sqrt{1 - \frac{a^2 c^2}{4 \omega^2}} \tag{2.281}$$

$$m_s = \frac{a\,A_0\,\rho_0\,c^2}{2\,\omega^2} \qquad (2.282)$$

R_s is the radiation resistance and m_s is the medium moving mass. Note that $R_s = 0$ when $a\,c/2\,\omega = 1$. The lower limit frequency of the exponential horn is obtained from the following formula.

$$f_0 = \frac{a\,c}{4\,\pi} \qquad (2.283)$$

The formula is only valid for an infinitely long exponential horn. If the wavelength is much smaller than the orifice diameter, the exponential horn can be considered as infinitely long.

Notations

Roman characters

A Projected area in the flow direction (m^2).
a Half of the width in a rectangular channel (m) or half of the width in a slit (m).
b Half of the height in a rectangular channel (m).
C Constant (m^2/s^2) or closed curve.
c The total concentration (mol/m^3) or the velocity of sound (m/s).
c_p Heat capacity at constant pressure (J/kg,K).
c_v Heat capacity at constant volume (J/kg,K).
D Length in the Reynolds number or diameter (m).
D_{ij} Deformation rate tensor (m/s).
F Absolute value of the vector (F_1, F_2, F_3) (N).
F_i Component i in the force vector (N).
f Friction factor.
f_i Component i in the vector (f_1, f_2, f_3).
G Mass flux $(kg/s, m^2)$.
g The absolute value of the vector (g_1, g_2, g_3) (m/s^2).
g_i Component i in the gravitational field (g_1, g_2, g_3) (m/s^2).
h Length coordinate, directed in the opposite direction to a homogeneous gravitational field (m).
M Torque (Nm).
N Molar flux (mol/s, m^2).
n_i Component i in the unit normal.
P Pressure function (m^2/s^2).
p The pressure of the fluid (N/m^2).
p' Dimensionless pressure $p' = p/\rho\,v_0^2$.
p_0 Pressure (N/m^2).
q Volume flow (m^3/s).

R Radius (m).

r Radial coordinate (m).

Re The Reynolds number.

S Closed boundary surface.

S' Closed boundary surface in the variable space (ξ_1, ξ_2, ξ_3).

s Length coordinate along a streamline (m).

t Time (s).

u The theoretical maximum velocity of the flow (m/s).

V Control volume enclosed by S.

v The absolute value of the velocity vector (v_1, v_2, v_3) (m/s).

$\mathbf{v}$ The velocity vector (v_1, v_2, v_3) (m/s).

v_i Component i in the velocity vector (v_1, v_2, v_3) (m/s).

v_i' Dimensionless velocity $v_i' = v_i/v_0$.

v_0 Velocity (m/s).

$\langle v \rangle$ Linear velocity (m/s).

x_i Component i in the space vector (x_1, x_2, x_3) (m).

Greek characters

Δp Sound pressure (N/m^2).

Δp_z Pressure difference (N/m^2).

ΔF_z Force (N).

Δz Length (m).

δ_{ij} The Kronecker delta. $\delta_{ij} = 1$ if $i = j$. $\delta_{ij} = 0$ if $i \neq j$.

ϵ_{ijk} The Levi-Civita symbol. $\epsilon_{ijk} = 1$ if i, j, k is an even permutation of $(1, 2, 3)$. $\epsilon_{ijk} = -1$ if i, j, k is an odd permutation of $(1, 2, 3)$. $\epsilon_{ijk} = 0$ if at least two indices are equal.

θ Angle coordinate in spherical coordinates.

κ Volume viscosity coefficient (kg/m,s).

μ First coefficient of viscosity (kg/m,s).

λ Second coefficient of viscosity (kg/m,s).

ξ_i Dimensionless space coordinate $\xi_i = x_i/D$.

ρ Density of the fluid (kg/m^3).

σ_{ij} Stress tensor (N/m^2).

τ_{ij} Viscous stress tensor (N/m^2).

Φ Velocity potential (m^2/s).

φ Angle coordinate in spherical coordinates.

Ω Potential function (m^2/s^2).

Chapter 3
Energy Transport

3.1 Introduction

This chapter describes the energy transport equations which are valid for a Newtonian fluid. Energy transport in a flowing Newtonian fluid is caused by thermal
conduction and convection. These equations are often written with tensor notation in
the chapter but sometimes even with vector notation. Note that the Einstein summation convention is valid for tensors. This means that summation occurs implicitly for
repeated indices. The derivative of a function f with respect to the space coordinates
x_i is denoted $f_{,i}$ with tensor notation. The chapter also includes some examples
of analytical solutions to the heat transport equation in some simple geometries.
Chapter 2 contains the solutions to the Navier-Stokes equation in these geometries.
These solutions have been used in the solution of the heat transport equation in this
chapter. See also the introduction in Chap. 2.

3.2 Energy Balance

The energy balance over a fixed closed surface S which encloses the volume V in a
Newtonian fluid can be written

$$\frac{\partial}{\partial t} \int_V \left(\rho \widehat{U} + \frac{1}{2} \rho v^2 + \rho \Omega \right) dV + \int_S \left(\rho \widehat{U} + \frac{1}{2} \rho v^2 + \rho \Omega \right) v_i n_i \, dS$$

$$+ \int_V \rho g_i v_i \, dV - \int_S \sigma_{ji} v_i n_j \, dS + \int_S Q_i n_i \, dS = 0 \qquad (3.1)$$

$$\sigma_{ji} = \sigma_{ij} = -p \, \delta_{ij} + \tau_{ij} \qquad (3.2)$$

The original version of this chapter was revised: For detailed information please see Erratum.
An erratum to this chapter can be found at https://doi.org/10.1007/978-3-319-01309-1_7.

P. Olsson, *Transport Phenomena in Newtonian Fluids - A Concise Primer*,
SpringerBriefs in Continuum Mechanics,
DOI 10.1007/978-3-319-01309-1_3, © The Author(s) 2014

$$\tau_{ij} = \lambda \, \delta_{ij} \, v_{k,k} + \mu \, (\, v_{i,j} + v_{j,i} \,) \qquad v_{i,j} + v_{j,i} = 2 \, D_{ij} \qquad (3.3)$$

$$g_i = - \Omega_{,i} \qquad (3.4)$$

Equation (3.1) is the energy balance expressed as an integral equation. The first term is the time derivative of the total energy in the volume V. The second term is the total convective energy flow through the surface S. The third term is the gravitational work done on the fluid in the volume V. The fourth term is the work being done on the surroundings outside the surface S. The fifth term is the conductive heat flow through the surface S. Equations (3.2) and (3.3) assumes that the fluid is Newtonian. D_{ij} is the deformation rate tensor. μ and λ are called the first and second coefficients of viscosity. Note that $\sigma_{ij} = \sigma_{ji}$ for a Newtonian fluid. Generally, the integrand in the fourth integral in Eq. (3.1) will be $\sigma_{ji} \, v_i \, n_j$. g_i is the gravitational field. Since the gravitational field is conservative, Eq. (3.4) is valid where Ω is the potential energy. Ω is assumed to be time independent. The Gauss theorem can be written

$$\int_S f_i \, n_i \, dS = \int_V f_{i,i} \, dV \qquad (3.5)$$

By applying the Gauss theorem, the continuity equation Eq. (2.15) and Eq. (3.4) on the third term in the integrand in the second integral in Eq. (3.1), the term can be written

$$\int_S \rho \, \Omega \, v_i \, n_i \, dS = - \int_V \rho \, g_i \, v_i \, dV - \int_V \Omega \, \frac{\partial \rho}{\partial t} \, dV \qquad (3.6)$$

If Ω is time independent, the time derivative of the third term in the integrand in the first integral can be written

$$\frac{\partial}{\partial t} \int_V \rho \, \Omega \, dV = \int_V \Omega \, \frac{\partial \rho}{\partial t} \, dV \qquad (3.7)$$

The third integral in Eq. (3.1) and the third term in the integrand in the first and second integral vanish. Then, the energy balance (3.1) can be written

$$\frac{\partial}{\partial t} \int_V \left(\rho \, \widehat{U} + \frac{1}{2} \rho \, v^2 \right) dV + \int_S \left(\rho \, \widehat{U} + \frac{1}{2} \rho \, v^2 \right) v_i \, n_i \, dS$$
$$- \int_S \sigma_{ji} \, v_i \, n_j \, dS + \int_S Q_i \, n_i \, dS = 0 \qquad (3.8)$$

This means that the energy balance is only influenced by the gravitational field through the Navier-Stokes-Duhem equation since the Navier-Stokes-Duhem equation contains the gravitational field g_i. The velocity vector v_i and the pressure p are influenced by the gravitational field. The enthalpy $\widehat{H}$ is defined

$$\rho \widehat{U} = \rho \widehat{H} - p \tag{3.9}$$

where $\widehat{U}$ is the internal energy and p is the pressure. Equations (3.2) and (3.9) give

$$\int_S \rho \widehat{U} v_i n_i \, dS - \int_S \sigma_{ij} v_i n_j \, dS = \int_S \rho \widehat{H} v_i n_i \, dS - \int_S \tau_{ij} v_i n_j \, dS \tag{3.10}$$

Then, Eq. (3.8) can be written

$$\frac{\partial}{\partial t} \int_V \left(\rho \widehat{H} + \frac{1}{2} \rho v^2 \right) dV + \int_S \left(\rho \widehat{H} + \frac{1}{2} \rho v^2 \right) v_i n_i \, dS$$
$$- \int_S \tau_{ij} v_i n_j \, dS + \int_S Q_i n_i \, dS = \frac{\partial}{\partial t} \int_V p \, dV \tag{3.11}$$

If $\partial p / \partial t$, ρv^2 and τ_{ij} are negligible, Eq. (3.11) can approximately be written

$$\frac{\partial}{\partial t} \int_V \rho \widehat{H} \, dV + \int_S \rho \widehat{H} v_i n_i \, dS + \int_S Q_i n_i \, dS = 0 \tag{3.12}$$

Equation (3.12) is the enthalpy balance expressed as an integral equation. By applying the Gauss theorem (3.5) on the surface integrals in Eq. (3.11), the equation can be written

$$\frac{\partial}{\partial t} \left(\rho \widehat{H} + \frac{1}{2} \rho v^2 \right) + \left[\left(\rho \widehat{H} + \frac{1}{2} \rho v^2 \right) v_i \right]_{,i}$$
$$- \left(\tau_{ij} v_i \right)_{,j} + Q_{i,i} = \frac{\partial p}{\partial t} \tag{3.13}$$

The continuity equation Eq. (2.15) means that Eq. (3.13) can be written

$$\rho \left[\frac{\partial}{\partial t} \left(\widehat{H} + \frac{1}{2} v^2 \right) + \left(\widehat{H} + \frac{1}{2} v^2 \right)_{,i} v_i \right]$$
$$- \left(\tau_{ij} v_i \right)_{,j} + Q_{i,i} = \frac{\partial p}{\partial t} \tag{3.14}$$

Equation (3.14) is the energy balance expressed as a differential equation. The term $(\tau_{ij} v_i)_{,j}$ in the equation represents the heat formation in the fluid caused by viscous friction. Equation (3.12) can in the same way be written

$$\rho \left(\frac{\partial \widehat{H}}{\partial t} + \widehat{H}_{,i} v_i \right) + Q_{i,i} = 0 \tag{3.15}$$

Equation (3.15) is the enthalpy balance expressed as a differential equation.

3.3 The Heat Transport Equation

The enthalpy $\widehat{H}$ can be written

$$\widehat{H} = \widehat{H}_0 + \Delta\widehat{H} \tag{3.16}$$

$$\Delta\widehat{H} = \int_{T_0}^{T} c_p(T')\,dT' \tag{3.17}$$

where $\widehat{H}_0$ is the enthalpy at the temperature T_0 and c_p is the heat capacity. If there occur chemical reactions in the fluid, $\widehat{H}_0$ will change with time in a point which follows the flow. This gives rise to a heat generation q which depends on the material time derivative of $\widehat{H}_0$ in the following way.

$$q + \left(\frac{\partial \widehat{H}_0}{\partial t} + \widehat{H}_{0,i}\, v_i \right) = 0 \tag{3.18}$$

The parenthesis in Eq. (3.18) is the material time derivative of $\widehat{H}_0$. Then, Eq. (3.14) can be written

$$\rho \left[\frac{\partial}{\partial t} \left(\Delta\widehat{H} + \frac{1}{2} v^2 \right) + \left(\Delta\widehat{H} + \frac{1}{2} v^2 \right)_{,i} v_i \right]$$
$$- q - \left(\tau_{ij}\, v_i \right)_{,j} + Q_{i,i} = \frac{\partial p}{\partial t} \tag{3.19}$$

The derivatives $\partial\Delta\widehat{H}/\partial t$ and $\Delta\widehat{H}_{,i}$ can be written

$$\frac{\partial\Delta\widehat{H}}{\partial t} = c_p \frac{\partial T}{\partial t} \tag{3.20}$$

$$\Delta\widehat{H}_{,i} = c_p\, T_{,i} \tag{3.21}$$

Then, Eq. (3.19) can be written

$$\rho c_p \left(\frac{\partial T}{\partial t} + T_{,i}\, v_i \right) + \frac{1}{2}\rho \left[\frac{\partial\left(v^2\right)}{\partial t} + \left(v^2\right)_{,i} v_i \right]$$
$$- q - \left(\tau_{ij}\, v_i \right)_{,j} + Q_{i,i} = \frac{\partial p}{\partial t} \tag{3.22}$$

Fourier's law can be written

$$Q_i = -\lambda_c\, T_{,i} \tag{3.23}$$

where λ_c is the thermal conductivity of the fluid. If λ_c is constant, Eq. (3.22) can be written

$$\lambda_c\, T_{,ii} + \frac{\partial p}{\partial t} + q + \left(\tau_{ij}\, v_i\right)_{,j} = \rho\, c_p \left(\frac{\partial T}{\partial t} + T_{,i}\, v_i\right)$$
$$+ \frac{1}{2}\, \rho \left[\frac{\partial\left(v^2\right)}{\partial t} + \left(v^2\right)_{,i}\, v_i\right] \tag{3.24}$$

Equation (3.24) is the heat transport equation when Fourier's law is valid. Equation (3.15) can, in the same way, be written

$$\lambda_c\, T_{,ii} + q = \rho\, c_p \left(\frac{\partial T}{\partial t} + T_{,i}\, v_i\right) \quad \text{or}$$
$$\lambda_c\, \nabla^2 T + q = \rho\, c_p \left(\frac{\partial T}{\partial t} + \nabla T \cdot \mathbf{v}\right) \tag{3.25}$$

In Eq. (3.25), the kinetic energy, the time derivative of the pressure and the friction forces are neglected. It is common that Eq. (3.25) is written

$$D_t\, \nabla^2 T + \frac{q}{\rho\, c_p} = \frac{\partial T}{\partial t} + \nabla T \cdot \mathbf{v} \tag{3.26}$$

$$D_t = \frac{\lambda_c}{\rho\, c_p} \tag{3.27}$$

where D_t is the thermal diffusivity.

3.4   Heat Transfer Coefficient

Consider a fixed object, bounded by the closed surface S, which is located in a flowing fluid in an infinite space where there are no other objects. Assume that the fluid temperature T_b is constant at infinite distance from the object and that the object temperature T_s is constant on the surface S. The heat flux Q_i on the surface S can be expressed by Fourier's law (3.23). The heat transfer coefficient α is defined

$$\int_S Q_i\, n_i\, dS = \alpha\, A\, (T_s - T_b) \tag{3.28}$$

where A is the area of the surface S and T_b is the *bulk temperature* of the fluid. The unit normal n_i is directed out from the surface S. Note that the heat flux Q_i in the general case does not need to be directed in the same direction as n_i.

Consider a straight channel with constant cross section which contains an incompressible flowing fluid with constant viscosity μ and heat capacity c_p. Assume that the flow profile v_i is independent of the axial coordinate x and the time t. Further, assume that S_x is a plane cross section which is perpendicular to the x-axis and bounded by the channel wall. The bulk temperature T_b is defined

$$\int\limits_{S_x} T \, v_i \, n_i \, dS = T_b \int\limits_{S_x} v_i \, n_i \, dS \qquad (3.29)$$

Consider a segment with the length Δx along the x-axis. Assume that ΔS is the surface of the channel in the segment and that the temperature T_s of the surface is constant on the boundary for constant x. Then, the heat transfer coefficient α between the surface of the channel and the fluid is defined

$$\lim_{\Delta A \to 0} \frac{1}{\Delta A} \int\limits_{\Delta S} Q_i \, n_i \, dS = \alpha \, (T_s - T_b) \qquad (3.30)$$

where ΔA is the area of the surface ΔS. The unit normal n_i is directed towards the center of the channel. In the both cases above, the bulk temperature T_b is defined in quite different way. If the space contains several objects, one can not define a bulk temperature unless the objects are far apart. The heat flow from a single object is influenced by the presence of the other objects. If the objects are grouped in a cluster, one can define the bulk temperature as the temperature of the fluid at infinite distance from the cluster. Then, it is possible to define a common heat transfer coefficient for the cluster if the objects have the same temperature. In this case, the cluster can be considered as a single object. Consider a large number of uniform objects which are regularly placed in the space in such way that the flow pattern around all objects becomes equal. This case is very similar to the heat transfer in a channel. One can use Eq. (3.29) to define the bulk temperature in this case if S_x is a cross-sectional surface perpendicular to the expectation value of the velocity vector. The heat transfer coefficient will depend somewhat on where the cross cross-sectional surface S_x is positioned along the x-axis. Then, one must define the heat transfer coefficient as an expectation value. In a packed bed of particles, the particles are randomly distributed in the space. One can even in this case define the bulk temperature approximately by Eq. (3.29). There are some limitations of using heat transfer coefficients.

Consider an incompressible fluid with constant viscosity μ, constant heat capacity c_p and constant thermal conductivity λ_c. Assume that the fluid flows around a solid object with a given geometry and that the flow is time independent. For uniform objects, the geometry of an object can be described by only one length quantity D. D can, for example, be the diameter of a sphere. Further, assume that the object is oriented in a given way relative to the flow direction and that the flow at infinite distance from the object is homogeneous with constant velocity v_0. Then, one can introduce dimensionless space coordinates ξ_i, dimensionless velocities v_i' and dimensionless temperature T' in the following way.

$$\xi_i = \frac{x_i}{D} \tag{3.31}$$

$$v_i' = \frac{v_i}{v_0} \tag{3.32}$$

$$T' = \frac{T - T_b}{T_s - T_b} \tag{3.33}$$

If $q = 0$, Eq. (3.25) can be written

$$\frac{\partial^2 T'}{\partial \xi_1^2} + \frac{\partial^2 T'}{\partial \xi_2^2} + \frac{\partial^2 T'}{\partial \xi_3^2} = Re \, Pr \left(v_1' \frac{\partial T'}{\partial \xi_1} + v_2' \frac{\partial T'}{\partial \xi_2} + v_3' \frac{\partial T'}{\partial \xi_3} \right) \tag{3.34}$$

$$Re = \frac{\rho \, v_0 \, D}{\mu} \tag{3.35}$$

$$Pr = \frac{\mu \, c_p}{\lambda_c} \tag{3.36}$$

where Re is the *Reynolds number* and Pr is the *Prandtl number*. The value of T' is equal to 1 on the surface of the object and equal to 0 at infinite distance from the object. This means that the boundary condition for T' is independent of T_s and T_b. Equation (2.72) implies that the flow pattern v_i' in the variable space (ξ_1, ξ_2, ξ_3) only depends on the Reynolds number Re for an object with a given shape. Then, Eq. (3.34) means that the temperature field T' in the variable space (ξ_1, ξ_2, ξ_3) only depends on Re and Pr for an object with a given shape. Fourier's law (3.23) means that Eq. (3.28) can be written

$$\alpha A \, (T_s - T_b) = -\lambda_c \int_S T_{,i} \, n_i \, dS$$

$$= -\lambda_c D \, (T_s - T_b) \int_{S'} \frac{\partial T'}{\partial \xi_i} \, n_i \, dS' \tag{3.37}$$

where S' is the surface of the object in the variable space (ξ_1, ξ_2, ξ_3) and $dS = D^2 dS'$. Equation (3.37) can also be written

$$\frac{\alpha A}{\lambda_c D} = - \int_{S'} \frac{\partial T'}{\partial \xi_i} \, n_i \, dS' \tag{3.38}$$

The left side in Eq. (3.38) depends only on Re and Pr. The area A of the outer surface S of the object is proportional to D^2 for an object with given shape. This means that the quantity $\alpha D / \lambda_c$ is a function of Re and Pr. The quantity $\alpha D / \lambda_c$ is called the *Nusselt number Nu*. Then, Eq. (3.38) can be written

$$\frac{\alpha D}{\lambda_c} = Nu(Re, Pr) \tag{3.39}$$

3.5   Analytical Solutions to the Heat Transport Equation

This section contains analytical solutions to the heat transport equation in some
simple geometries. These solutions are based on the analytical solutions to the
Navier-Stokes equation which are described in Chap. 2. The section also contains
some correlations for heat transfer in cases where there are no analytical solutions.

3.5.1   Heat Transfer Around a Sphere

The Navier-Stokes equation can be solved analytically for creeping flow around
a sphere. This solution is described in Sect. 2.7.1. The solution is only valid for
Reynolds numbers which are less than 0.1. At such low values of the Reynolds
number, the flow around the sphere will have little effect on the heat transfer. Therefore, it is not particularly meaningful to try to find an analytical solution for the temperature field around a sphere in the flow pattern which is described in Sect. 2.7.1.
It is only meaningful to solve the heat transport equation analytically for stagnant
flow in this case. At stationary state and stagnant flow in angle independent spherical
coordinates, Eq. (3.25) can be written in the following way if $q = 0$.

$$\frac{\partial^2}{\partial r^2} (r\, T) = 0 \tag{3.40}$$

The solution can be written

$$T = T_b + (T_s - T_b)\, \frac{R}{r} \tag{3.41}$$

where T_s is the surface temperature of the sphere, T_b is the temperature of the fluid
at infinite distance from the sphere and R is the radius of the sphere. The heat flux
Q_r on the surface of the sphere can be written

$$Q_r = \lambda_c\, (T_s - T_b)/R = \alpha\, (T_s - T_b) \tag{3.42}$$

If D is the diameter of the sphere, the Nusselt number $Nu = \alpha\, D/\lambda_c$ can be written

$$Nu = 2 \tag{3.43}$$

According to Eq. (3.39), Nu is a function of Re and Pr. The following correlation
is taken from *Transport Phenomena by Beek & Muttzall* [1].

$$Nu = 2 + 1.3 \cdot Pr^{0.15} + 0.66 \cdot Re^{0.50} \cdot Pr^{0.33} \qquad 1 < Re < 10^4 \tag{3.44}$$

The following correlation is taken from *Transport Phenomena by Bird, Stewart & Lightfoot* [2].

$$Nu = 2 + 0.60 \cdot Re^{1/2} \cdot Pr^{1/3} \qquad 1 < Re < 5 \cdot 10^4 \tag{3.45}$$

In both cases, Re and Pr are defined according to Eq. (3.35) and (3.36) where D is the diameter of the sphere. Q_r in Eq. (3.42) is not constant on the surface of the sphere for non-stagnant flow. Assume that the flow is directed in the z-axis direction and that the center of the sphere coincides with the origin. See Fig. 2.1. Then, Q_θ is nonzero. In the general case, α is defined by Eq. (3.28).

3.5.2 Heat Transfer in a Cylindrical Tube

Consider a cylindrical tube which contains a flowing incompressible fluid with constant viscosity μ, constant thermal conductivity λ_c and constant heat capacity c_p. Assume that the flow pattern is independent of the time and the axial coordinate z in a cylindrical coordinate system and that the center line of the cylinder coincides with the z-axis. Equation (2.93) is the solution to the flow problem. If $q = 0$, Eq. (3.25) can be written

$$\lambda_c \left(\frac{\partial^2 T}{\partial r^2} + \frac{1}{r} \frac{\partial T}{\partial r} + \frac{\partial^2 T}{\partial z^2} \right) = \rho\, c_p\, v_0 \left(1 - \frac{r^2}{R^2} \right) \frac{\partial T}{\partial z} \tag{3.46}$$

where R is the radius of the cylinder and v_0 is the velocity of the fluid in the center of the cylinder. Equation (3.46) is called the *Graetz equation*. In this case, the bulk temperature T_b is defined by Eq. (3.29). T_b varies with the axial coordinate z in a way which depends on the boundary condition on the tube wall. The radial temperature dependence also depends on the boundary condition at the tube wall. This section describes the case when the radial heat flux at the tube wall is constant. In this case, $\partial^2 T/\partial z^2 = 0$ and $\partial T/\partial z$ is constant in Eq. (3.46). For other types of boundary conditions, Eq. (3.46) must be solved from case to case to obtain an exact solution. If the radial heat flux at the tube wall is constant and independent of z, the bulk temperature T_b will vary linearly with z. This is also valid for the temperature field $T(r, z)$ which can be written

$$T(r, z) = T_r(r) + a\,z \tag{3.47}$$

where $T_r(r)$ is a function which only depends on r and a is a constant independent of r and z. Then, Eq. (3.46) can be written

$$\lambda_c \left(\frac{\partial^2 T_r}{\partial r^2} + \frac{1}{r} \frac{\partial T_r}{\partial r} \right) = a\,\rho\, c_p\, v_0 \left(1 - \frac{r^2}{R^2} \right) \tag{3.48}$$

The solution to Eq. (3.48) can be written

$$T_r = \frac{a \rho c_p v_0}{4 \lambda_c} r^2 \left(1 - \frac{r^2}{4 R^2}\right) + T_c \tag{3.49}$$

where T_c is a constant independent of r and z. The radial heat flux $Q_r(r)$ can be written

$$Q_r(r) = -\lambda_c \frac{\partial T_r}{\partial r} = -\frac{1}{4} a \rho c_p v_0 r \left(2 - \frac{r^2}{R^2}\right) \tag{3.50}$$

The radial heat flux at the tube wall can be written

$$Q_r(R) = -\frac{1}{4} a \rho c_p v_0 R = -\alpha (T_s - T_b) \tag{3.51}$$

where α is the heat transfer coefficient and T_s is the temperature of the tube wall. The minus sign in front of α in Eq. (3.51) is a consequence of that the unit normal n_i in Eq. (3.30) is directed towards the tube center. The constant T_c in Eq. (3.49) is determined by the boundary condition $T(R, z) = T_s$. Then, $T(r, z)$ can be written

$$T(r, z) = T_s + \frac{a \rho c_p v_0}{4 \lambda} \left[r^2 \left(1 - \frac{r^2}{4 R^2}\right) - \frac{3}{4} R^2\right] \tag{3.52}$$

Equations (3.29), (2.93) and (3.52) means that the bulk temperature T_b can be written

$$T_b = T_s - \frac{11}{96} \frac{a \rho c_p v_0}{\lambda_c} R^2 \tag{3.53}$$

Insertion of Eq. (3.53) into Eq. (3.51) gives

$$\alpha = \frac{24}{11} \frac{\lambda_c}{R} \tag{3.54}$$

The Nusselt number Nu for a cylindrical tube is defined

$$Nu = \frac{\alpha D}{\lambda_c} \tag{3.55}$$

where $D = 2R$. Then, Eq. (3.54) can be written

$$Nu = \frac{48}{11} \approx 4.36 \tag{3.56}$$

For fully developed laminar flow in a tube, there are no acceleration forces in the flow. This means that Nu will be independent of Re and Pr for constant heat flux at the tube wall. Equation (3.56) is valid for constant heat flux at the tube wall. In this case, $\partial^2 T/\partial z^2 = 0$ and $\partial T/\partial z = a$ in Eq. (3.46) where a is a constant. For other types of boundary conditions, both $\partial^2 T/\partial z^2$ and $\partial T/\partial z$ are functions of r and z and v_0 is a parameter in these functions. This means that the local heat transfer coefficient α depends somewhat on z and the flow in the tube in this case. It is only for constant heat flux on the tube wall that the local heat transfer coefficient is completely independent of z and the flow in the tube. There are reports in the literature that $Nu = 48/11$ for constant heat flux on the tube wall and $Nu = 3.66$ for constant wall temperature. The latter value of Nu is approximately valid if the bulk temperature T_b varies moderately with z.

The following correlations are taken from *Transport Phenomena by Bird, Stewart & Lightfoot* [3].

$$Nu = 7 + 0.025 \, (Re \, Pr)^{0.8} \tag{3.57}$$

$$Nu = 5 + 0.025 \, (Re \, Pr)^{0.8} \tag{3.58}$$

The both correlations are valid for turbulent flow. Equation (3.57) is valid for constant heat flux on the tube wall and Eq. (3.58) is valid for constant temperature on the tube wall. Transition to turbulent flow occurs when $Re \approx 2100$.

3.5.3 Heat Transfer in a Rectangular Channel

Consider a rectangular channel with the width $2a$ and the height $2b$ which contains a flowing incompressible fluid with constant viscosity μ, constant thermal conductivity λ_c and constant heat capacity c_p. Assume that the channel center line coincides with the z-axis in in a Cartesian coordinate system as in Fig. 2.3. Assume that the flow is stationary and independent of the coordinate z. This is valid for fully developed flow profile. Equation (2.121) is the solution to the flow problem. If $q = 0$, Eq. (3.25) can be written

$$
\lambda_c \left(\frac{\partial^2 T}{\partial x^2} + \frac{\partial^2 T}{\partial y^2} + \frac{\partial^2 T}{\partial z^2} \right)
$$

$$
= \rho \, c_p \frac{\partial T}{\partial z} \sum_{n=0}^{\infty} \sum_{m=0}^{\infty} v_{nm} \cos\left[\frac{\pi x}{2a}(1 + 2n) \right] \cos\left[\frac{\pi y}{2b}(1 + 2m) \right] \tag{3.59}
$$

where v_{nm} is obtained from Eq. (2.120). In this case, the bulk temperature T_b is defined by Eq. (3.29). T_b varies with the axial coordinate z in a way which depends on the type of boundary condition which is valid at the channel wall. The x- and y-dependence of the temperature also depends on the boundary condition at the channel wall. This section describes the case when the heat formation on the channel wall is constant while the temperature on the channel wall is constant in a cross-section of the channel. This boundary condition can approximately be realized if the heat formation on the channel surface is constant while the thermal conductivity in the solid material around the channel is high enough. The thermal conductivity must not be so high that the axial heat conduction in the solid material can not be neglected. This is a quite realistic situation for long and narrow channels. In this case, the heat flux depends on x and y in a cross section on the channel surface but is independent of z for constant x and y on the channel surface. If the heat formation on the channel wall is constant and independent of z, the bulk temperature T_b will vary linearly with z. This also applies to the temperature field $T(x, y, z)$ which can be written

$$T(x, y, z) = T_{xy}(x, y) + T_c + c_z z \tag{3.60}$$

where $T_{xy}(x, y)$ is a function which only depends on x and y and T_c and c_z are constants which are independent of x, y and z. Then, Eq. (3.59) can be written

$$\frac{\partial^2 T_{xy}}{\partial x^2} + \frac{\partial^2 T_{xy}}{\partial y^2}$$
$$= \frac{c_z \rho c_p}{\lambda_c} \sum_{n=0}^{\infty} \sum_{m=0}^{\infty} v_{nm} \cos\left[\frac{\pi x}{2a}(1 + 2n)\right] \cos\left[\frac{\pi y}{2b}(1 + 2m)\right] \tag{3.61}$$

For constant temperature in a cross section on the channel surface, $T_{xy}(x, y)$ can be written

$$T_{xy}(x, y) = \sum_{n=0}^{\infty} \sum_{m=0}^{\infty} T_{nm} \cos\left[\frac{\pi x}{2a}(1 + 2n)\right] \cos\left[\frac{\pi y}{2b}(1 + 2m)\right] \tag{3.62}$$

Insertion of Eqs. (2.120) and (3.62) into Eq. (3.61) means that T_{nm} can be written

$$T_{nm} = -\frac{c_z \rho c_p}{\lambda_c} \cdot$$
$$4\left(\frac{2}{\pi}\right)^6 \frac{a^2 b^2 C (-1)^{n+m}}{(1 + 2n)(1 + 2m)\left[\frac{b}{a}(1 + 2n)^2 + \frac{a}{b}(1 + 2m)^2\right]^2} \tag{3.63}$$

$$C = \frac{\Delta p}{\mu \Delta z} \tag{3.64}$$

See also Eqs. (2.105) and (2.122). Equation (3.60) is the solution to Eq. (3.59) when the heat formation on the channel surface is constant and independent of x, y and z and the thermal conductivity in the solid material is so high that one can assume that the surface temperature is constant in a cross section in the channel. The sum in Eq. (3.62) is equal to zero on the channel surface which means that the temperature T_s on the surface is independent of x and y but varies linearly with z. The constant T_c in Eq. (3.60) is the surface temperature when $z = 0$. The heat flux on the surface is independent of q but varies with x and y. Equations (3.23), (3.28), (3.60), (3.64), (2.122) and (2.123) give the following equation.

$$\frac{c_z \rho \, c_p \, q}{4\,(a+b)} = \alpha \,(T_s - T_b) \tag{3.65}$$

q is the volume flow in the channel and T_s is the surface temperature on the channel wall. See Eq. (2.122). Equation (3.65) is consistent with the assumption that the bulk temperature T_b varies linearly with z in such a way that $\partial T_b/\partial z = c_z$. The surface temperature T_s can be written

$$T_s = T_c + c_z z \tag{3.66}$$

The bulk temperature T_b is obtained from Eq. (3.29) which can be written

$$q \, T_b = \int_{-a}^{a} \int_{-b}^{b} \left[T_s + T_{xy}(x,y) \right] v_z(x,y) \, dx \, dy$$

$$= q \, T_s + \int_{-a}^{a} \int_{-b}^{b} T_{xy}(x,y) \, v_z(x,y) \, dx \, dy \tag{3.67}$$

where $v_z(x,y)$ is obtained from Eq. (2.121). The heat transfer coefficient α is obtained from Eqs. (3.65) and (3.67). The integrand in the double integral in Eq. (3.67) can be written

$$T_{xy}(x,y) \, v_z(x,y) = \sum_{n=0}^{\infty} \sum_{m=0}^{\infty} \sum_{l=0}^{\infty} \sum_{k=0}^{\infty} T_{nm} \, v_{lk}$$

$$\cdot \cos\left[\frac{\pi x}{2a}(1+2n) \right] \cos\left[\frac{\pi x}{2a}(1+2l) \right]$$

$$\cdot \cos\left[\frac{\pi y}{2b}(1+2m) \right] \cos\left[\frac{\pi y}{2b}(1+2k) \right] \tag{3.68}$$

Each term in the sum in Eq. (3.68) is a product between a function which depends on x and a function which depends on y. Since the limits in the double integral in Eq. (3.67) are constant, the integral can be written

$$\int_{-a}^{a} \int_{-b}^{b} T_{xy}(x,y) \, v_z(x,y) \, dx \, dy = \sum_{n=0}^{\infty} \sum_{m=0}^{\infty} \sum_{l=0}^{\infty} \sum_{k=0}^{\infty} T_{nm} \, v_{lk}$$

$$\cdot \int_{-a}^{a} \cos\left[\frac{\pi x}{2a}(1+2n) \right] \cos\left[\frac{\pi x}{2a}(1+2l) \right] dx$$

$$\cdot \int_{-b}^{b} \cos\left[\frac{\pi y}{2b}(1+2m) \right] \cos\left[\frac{\pi y}{2b}(1+2k) \right] dy \tag{3.69}$$

The integrands in the integrals in Eq. (3.69) can be written

$$
\cos\left[\frac{\pi x}{2a}(1+2n)\right]\cos\left[\frac{\pi x}{2a}(1+2l)\right]
$$

$$
= \frac{1}{2}\cos\left[\frac{\pi x}{a}(n-l)\right] + \frac{1}{2}\cos\left[\frac{\pi x}{a}(1+n+l)\right] \tag{3.70}
$$

$$
\cos\left[\frac{\pi y}{2b}(1+2m)\right]\cos\left[\frac{\pi y}{2b}(1+2k)\right]
$$

$$
= \frac{1}{2}\cos\left[\frac{\pi y}{b}(m-k)\right] + \frac{1}{2}\cos\left[\frac{\pi y}{b}(1+m+k)\right] \tag{3.71}
$$

Then, the integrals in Eq. (3.69) can be written

$$
\int_{-a}^{a}\cos\left[\frac{\pi x}{2a}(1+2n)\right]\cos\left[\frac{\pi x}{2a}(1+2l)\right]dx = \begin{cases} a & \text{if } n = l \\ 0 & \text{if } n \neq l \end{cases} \tag{3.72}
$$

$$
\int_{-b}^{b}\cos\left[\frac{\pi y}{2b}(1+2m)\right]\cos\left[\frac{\pi y}{2b}(1+2k)\right]dy = \begin{cases} b & \text{if } m = k \\ 0 & \text{if } m \neq k \end{cases} \tag{3.73}
$$

Equations (3.72) and (3.73) means that Eq. (3.69) can be written

$$
\int_{-a}^{a}\int_{-b}^{b} T_{xy}(x,y)\, v_z(x,y)\, dxdy = ab\sum_{n=0}^{\infty}\sum_{m=0}^{\infty} T_{nm}\, v_{nm} \tag{3.74}
$$

Equations (2.120), (2.122), (2.123), (3.63), (3.64), (3.65), (3.67) and (3.74) means that the heat transfer coefficient α can be written

$$
\alpha = \left(\frac{8}{\pi}\right)^2 \frac{\Sigma^2}{\Sigma_T} \frac{\lambda_c}{4(a+b)} \tag{3.75}
$$

$$
\Sigma_T = \sum_{n=0}^{\infty}\sum_{m=0}^{\infty} \frac{1}{(1+2n)^2(1+2m)^2\left[\frac{b}{a}(1+2n)^2 + \frac{a}{b}(1+2m)^2\right]^3} \tag{3.76}
$$

where Σ is obtained from Eq. (2.123). The hydraulic diameter D_h for a rectangular channel with the width $2a$ and the height $2b$ is obtained from Eq. (2.124). The Nusselt number Nu is defined

$$
Nu = \frac{\alpha D_h}{\lambda_c} = \left(\frac{8}{\pi}\right)^2 \frac{\Sigma^2}{\Sigma_T} \frac{ab}{(a+b)^2} \tag{3.77}
$$

Table 3.1 Nusselt number
Nu in a rectangular channel
as function of a/b

a/b	Nu
1.0	3.608
0.9	3.620
0.8	3.664
0.7	3.750
0.6	3.895
0.5	4.123
0.4	4.472
0.3	4.990
0.2	5.738
0.1	6.785
0.0	8.235

Table 3.1 shows Nu for different values of a/b. Equation (3.77) is valid for fully developed laminar flow and constant heat formation on the channel surface when the thermal conductivity in the solid material around the channel is so high that one can assume that the temperature on the channel surface in a cross section in the channel is constant and independent of x and y.

3.5.4   Heat Transfer in a Slit

Consider a slit with the width $2a$ which contains a flowing incompressible fluid with constant viscosity μ, constant thermal conductivity λ_c and constant heat capacity c_p. Assume that the center surface of the slit coincides with the y-z-plane in a Cartesian coordinate system as in Fig. 2.4 and that the slit has infinite extension in the y-z-plane. Further, assume that the flow is stationary and independent of the coordinates y and z and that the flow is directed in the z coordinate direction. Equation (2.133) is the solution to the flow problem. If $q = 0$, Eq. (3.25) can be written

$$\lambda_c \left(\frac{\partial^2 T}{\partial x^2} + \frac{\partial^2 T}{\partial z^2} \right) = \rho \, c_p \, v_0 \left(1 - \frac{x^2}{a^2} \right) \frac{\partial T}{\partial z} \tag{3.78}$$

The bulk temperature T_b is defined by Eq. (3.29). T_b varies with the coordinate z in a way which depends on the boundary condition on the slit wall. The x dependence also depends on the boundary condition on the wall. This section describes the case when the heat flux on the slit wall is constant. In this case, $\partial^2 T/\partial z^2 = 0$ and $\partial T/\partial z$ is a constant in Eq. (3.78). For other types of boundary conditions, Eq. (3.78) must be solved from case to case to obtain an exact solution. If the heat flux on the slit wall is constant and independent of y and z, the bulk temperature T_b will vary linearly with z. This is also valid for the temperature field $T(x, z)$ which can be written

$$T(x, z) = T_x(x) + c_z z \tag{3.79}$$

where $T_x(x)$ is a function which only depends on x and c_z is a constant which is independent of x and z. Then, Eq. (3.78) can be written

$$\frac{\partial^2 T_x}{\partial x^2} = \frac{c_z \rho c_p v_0}{\lambda_c} \left(1 - \frac{x^2}{a^2}\right) \tag{3.80}$$

The solution to Eq. (3.80) can be written

$$T_x = \frac{c_z \rho c_p v_0}{2 \lambda_c} x^2 \left(1 - \frac{x^2}{6 a^2}\right) + T_c \tag{3.81}$$

where T_c is a constant, independent of x and z. The heat flux $Q_x(x)$ in the x coordinate direction can be written

$$Q_x(x) = -\lambda_c \frac{\partial T_x}{\partial x} = -c_z \rho c_p v_0 x \left(1 - \frac{x^2}{3 a^2}\right) \tag{3.82}$$

The heat flux at $x = a$ can be written

$$Q_x(a) = -\frac{2}{3} c_z \rho c_p v_0 a \tag{3.83}$$

The heat flux $Q_n = -Q_x(a) = Q_x(-a)$ between the boundary surfaces and the fluid can be written

$$Q_n = \frac{2}{3} c_z \rho c_p v_0 a = \alpha (T_s - T_b) \tag{3.84}$$

where α is the heat transfer coefficient between the slit surfaces and the fluid and T_s is the surface temperature. The constant T_c in Eq. (3.81) is determined by the boundary condition $T(a, z) = T_s$. Then, $T(x, z)$ can be written

$$T(x, z) = T_s + \frac{c_z \rho c_p v_0}{2 \lambda_c} \left[x^2 \left(1 - \frac{x^2}{6 a^2}\right) - \frac{5}{6} a^2 \right] \tag{3.85}$$

The bulk temperature T_b is obtained from Eq. (3.29) which can be written

$$\int_{-a}^{a} T(x, z) v_z(x) \, dx = T_b \int_{-a}^{a} v_z(x) \, dx \tag{3.86}$$

Equations (2.133), (3.84), (3.85) and (3.86) means that the heat transfer coefficient α can be written

$$\alpha = \frac{35}{17} \frac{\lambda_c}{a} \tag{3.87}$$

According to Eq. (2.136), the hydraulic diameter D_h is equal to $4a$ for an infinite slit with the width $2a$. Then, the Nusselt number Nu can be written

$$Nu = \alpha \frac{4a}{\lambda_c} = \frac{140}{17} \approx 8.235 \tag{3.88}$$

This value is consistent with the value of Nu in Table 3.1 when $a/b = 0$. Equation (3.88) is only valid for fully developed laminar flow. For fully developed laminar flow in a slit, there are no acceleration forces in the flow. This means that Nu will be independent of Re and Pr for constant heat flux on the slit wall. Equation (3.88) is valid for constant heat flux on the wall. In this case, $\partial^2 T/\partial z^2 = 0$ and $\partial T/\partial z = c_z$ in Eq. (3.78) where c_z is a constant. For other types of boundary conditions, both $\partial^2 T/\partial z^2$ and $\partial T/\partial z$ in Eq. (3.78) are functions of x and z and v_0 is a parameter in these functions. This means that the local heat transfer coefficient α depends somewhat on z and the flow in the slit in this case. It is only for constant heat flux on the wall that the local heat transfer coefficient is completely independent of z and the flow in the slit. There are reports in the literature that $Nu = 140/17$ for constant heat flux and $Nu = 7.54$ for constant wall temperature on the slit wall. The latter value of Nu is approximately valid if the bulk temperature T_b varies moderately with z.

3.6 Empirical Correlations for Heat Transfer

3.6.1 Heat Transfer in a Packed Bed of Particles

The bulk temperature of an incompressible fluid in a packed bed of particles can approximately be defined by Eq. (3.29). See Sect. 3.4. The Nusselt number Nu is a function of the Reynolds number Re and the Prandtl number Pr. See Eq. (3.39). This function can be determined by fitting some empirical expression to experimental data. The following correlation is taken from *Transport Phenomena by Bird, Stewart & Lightfoot* [4].

$$Nu = j_H \, Re \, Pr^{1/3} \tag{3.89}$$

$$j_H = 0.91 \, Re^{-0.51} \, \Phi \quad (Re < 50) \tag{3.90}$$

$$j_H = 0.61 \, Re^{-0.41} \, \Phi \quad (Re > 50) \tag{3.91}$$

$$Re = \frac{G}{S_v \, \mu \, \Phi} \tag{3.92}$$

$$Pr = \frac{c_p \, \mu}{\lambda_c} \tag{3.93}$$

G is the superficial mass flow, S_v is the specific area of the bed, λ_c is the thermal conductivity of the fluid and Φ is the shape factor of the particles. Table 3.2 shows the value of Φ for different particle shapes. Colburn's j_H-factor is defined

Table 3.2 The shape factor
Φ for different particle shapes

Particle shape	Φ
Spheres	1.00
Cylinders	0.91
Flakes	0.86
Raschig rings	0.79
Partition rings	0.67
Berl-saddles	0.80

$$j_H = \frac{\alpha}{c_p G}\, Pr^{2/3} \tag{3.94}$$

where α is the heat transfer coefficient. The heat transfer coefficient α is obtained from the following equation.

$$\alpha = Nu\, S_v \lambda_c\, \Phi \tag{3.95}$$

Creeping flow in a packed bed means that the flow is so low that the acceleration forces in the fluid are negligible compared with the viscous forces. For creeping flow in a packed bed ($Re \lesssim 0.1$), the Nusselt number Nu is independent of Re. This means that the correlation for Nu above is not valid for very low values of Re. For creeping flow, $j_H \propto Re^{-1}$.

3.6.2 Heat Transfer in a Package of Wire Screens

The bulk temperature in a package of wire screens can approximately be defined by Eq. (3.29). See Sect. 3.4. The Nusselt number Nu is a function of the Reynolds number Re and the Prandtl number Pr. See Eq. (3.39). This function can be determined by fitting some empirical correlation to experimental data. The following correlation is taken from an article by *Satterfield & Cortez* [5].

$$Nu = j_H\, Re\, Pr^{1/3} \tag{3.96}$$

$$j_H = \frac{0.874}{\varepsilon_b}\left[4\left(1 - \varepsilon_b\right) Re\right]^{-0.645} \tag{3.97}$$

$$Re = \frac{G}{S_v\, \mu} \tag{3.98}$$

$$Pr = \frac{c_p\, \mu}{\lambda_c} \tag{3.99}$$

ε_b is the porosity of the bed. The correlation above assumes that the wire screens are placed perpendicular to the flow. The heat transfer coefficient α is obtained from the following equation.

$$\alpha = Nu \, S_v \lambda_c \tag{3.100}$$

Moreover, the same notations and equations are valid as in Sect. 3.6.1. See also Sect. 3.6.1.

Notations

Roman characters

A Area (m^2).

a Half of the width in a rectangular channel (m) or half of the width in a slit (m) or a constant in Eq. (3.47) (K/m).

b Half of the height in a rectangular channel (m).

c_p The heat capacity of the fluid (J/kg, K).

c_z Constant in Eq. (3.60) (K/m).

D Characteristic length (m).

D_h Hydraulic diameter (m).

D_{ij} The deformation rate tensor (m/s).

D_t Thermal diffusivity (m^2/s).

f_i Component i in the vector (f_1, f_2, f_3).

G Superficial mass flow (kg/m^2, s).

g_i Component i in the gravitational field (g_1, g_2, g_3) (m/s^2).

$\widehat{H}$ Enthalpy (J/kg).

$\widehat{H}_0$ Enthalpy at the temperature T_0 (J/kg).

j_H Colburn's j_H-factor.

Nu The Nusselt number.

n_i Component i in the unit normal vector.

Pr The Prandtl number.

p The pressure of the fluid (N/m^2).

Q_i Component i of the heat flux (Q_1, Q_2, Q_3) (J/s, m^2).

Q_x Heat flux (W/m^2).

Q_n Heat flux (W/m^2).

q Heat source (J/s, kg) or volume flow (m^3/s).

R Radius (m).

Re The Reynolds number.

r Radial coordinate (m).

S Closed boundary surface.

S_v Specific area (1/m).

T Absolute temperature (K).

T_b The bulk temperature of the fluid (K).

T_c Constant in Eq. (3.60) (K).

T_r Radial temperature function (K).

T_s Surface temperature (K).

T' Dimensionless temperature.

t Time (s).

$\hat{U}$ Internal energy (J/kg).

V Control volume enclosed by S.

v Absolute value of the velocity vector (v_1, v_2, v_3) (m/s).

$\mathbf{v}$ The velocity vector (v_1, v_2, v_3) (m/s).

v_0 The velocity in the middle of a channel (m/s).

v_i Component i in the velocity vector (v_1, v_2, v_3) (m/s).

v_i' Component i in the dimensionless velocity vector.

x Space coordinate (m).

y Space coordinate (m).

z Space coordinate (m).

Greek characters

α Heat transfer coefficient (W/m^2, K).

δ_{ij} The Kronecker delta. $\delta_{ij} = 1$ if $i = j$. $\delta_{ij} = 0$ if $i \neq j$.

ε_b Porosity in a packed bed.

μ First coefficient of viscosity of the fluid (kg/m, s).

λ_c Thermal conductivity of the fluid (W/m, K).

λ Second coefficient of viscosity of the fluid (kg/m, s).

ξ_i Dimensionless space coordinate.

ρ The density of the fluid (kg/m^3).

σ_{ij} The stress tensor (N/m^2).

τ_{ij} Viscous stress tensor (N/m^2).

Φ The shape factor of the particles.

Ω Potential function (J/kg).

References

1. W.J. Beek, K.M.K. Muttzall, *Transport Phenomena* (John Wiley and Sons Ltd, New York, 1975), p. 196, equation (III.68)
2. R.B. Bird, W.E. Stewart, E.N. Lightfoot, *Transport Phenomena* (John Wiley and Sons, New York, 1960), p. 409, equation (13.3-1)
3. R.B. Bird, W. E. Stewart, E.N. Lightfoot, *Transport Phenomena* (John Wiley and Sons, New York, 1960), p. 404, equation (13.2-21) and (13.2-22)
4. R.B. Bird, W.E. Stewart, E.N. Lightfoot, *Transport Phenomena* (John Wiley & Sons, New York, 1960), p. 411–412, equation (13.4-1) to (13.4-5)
5. C.N. Satterfield, D.H. Cortez, Mass transfer characteristics of woven-wire screen catalysts. Ind. Eng. Chem. Fundam. **9**(4), 613 (1970)

Chapter 4
Mass Transport

4.1 Introduction

This chapter describes the mass transport equations which are valid for a Newtonian fluid. Mass transport in a flowing Newtonian fluid is caused by diffusion and
convection. These equations are always written with vector notation in this chapter.
There is no tensor notation in this chapter. Index i and j are compound indices in
the chapter. The chapter also contains some examples of analytical solutions to the
mass transport equation for different geometries. The solutions to the flow equation
for these geometries are described in Chap. 2. These solutions have been used in the
solution of the mass transport equation in this chapter. See also the introduction in
Chap. 2.

4.2 Molar Balance

The molar balance over a fixed closed surface S which encloses the volume V in a
Newtonian fluid can be written

$$\frac{\partial}{\partial t} \int_V c_i \, dV + \int_S \mathbf{N}_i \cdot d\mathbf{S} = \int_V q_i \, dV \tag{4.1}$$

where c_i is the concentration of compound i, $\mathbf{N}_i$ is the molar flux of compound i and
q_i is the formation rate of compound i. By applying Gauss theorem on the surface
integral in Eq. (4.1), the equation can be written

The original version of this chapter was revised: For detailed information please see Erratum.
An erratum to this chapter can be found at https://doi.org/10.1007/978-3-319-01309-1_7.

P. Olsson, *Transport Phenomena in Newtonian Fluids - A Concise Primer,*
SpringerBriefs in Continuum Mechanics,
DOI 10.1007/978-3-319-01309-1_4, © The Author(s) 2014

$$\frac{\partial c_i}{\partial t} + \nabla \cdot \mathbf{N}_i = q_i \tag{4.2}$$

Equation (4.2) is the molar balance expressed as a differential equation.

4.3 Binary Diffusion

The binary diffusion coefficients $\mathscr{D}_{ij}$ and $\mathscr{D}_{ji}$ in a binary mixture of compound i and j are defined by Fick's first law which can be written

$$\mathbf{J}_i = -c\,\mathscr{D}_{ij}\nabla x_i \tag{4.3}$$
$$\mathbf{J}_j = -c\,\mathscr{D}_{ji}\nabla x_j \tag{4.4}$$

where $\mathbf{J}_i$ is the molar flux of compound i caused by diffusion, c is the total concentration and x_i is the molar fraction of compound i. It is common that Fick's first law is written

$$\mathbf{J}_i = -D_{ij}\nabla c_i \tag{4.5}$$
$$\mathbf{J}_j = -D_{ji}\nabla c_j \tag{4.6}$$

where $c_i = c\,x_i$. It is obvious that $D_{ij} \neq \mathscr{D}_{ij}$ if c depends on the space coordinates. $\mathbf{J}_i$ and $\mathbf{J}_j$ are defined by

$$\mathbf{N}_i = \mathbf{J}_i + x_i\left(\mathbf{N}_i + \mathbf{N}_j\right) \tag{4.7}$$
$$\mathbf{N}_j = \mathbf{J}_j + x_j\left(\mathbf{N}_i + \mathbf{N}_j\right) \tag{4.8}$$

where $\mathbf{N}_i$ is the molar flux of compound i caused by convection and diffusion. The equation $x_i + x_j = 1$ and summation of Eqs. (4.7) and (4.8) gives

$$\nabla x_i + \nabla x_j = 0 \tag{4.9}$$
$$\mathbf{J}_i + \mathbf{J}_j = 0 \tag{4.10}$$

Equations (4.3), (4.4), (4.9) and (4.10) give

$$\mathscr{D}_{ij} = \mathscr{D}_{ji} \tag{4.11}$$

In the general case, $\mathscr{D}_{ij}$ depends on the composition x_i, but in dilute gases, $\mathscr{D}_{ij}$ is nearly constant and independent of the composition.

4.4   Diffusion in Multi Component Systems of Gases

In a multi component system, the molar flux $\mathbf{J}_i$ is defined by

$$\mathbf{N}_i = \mathbf{J}_i + x_i \mathbf{N} \tag{4.12}$$

$$\mathbf{N} = \sum_{j=1}^{n} \mathbf{N}_j \tag{4.13}$$

where n is the number of compounds in the system and x_i is the molar fraction of compound i. $\mathbf{J}_i$ is the molar flux of compound i caused by diffusion. $\mathbf{N}_i$ is the molar flux of compound i caused by convection and diffusion. Summation of Eq. (4.12) for all i gives

$$\sum_{i=1}^{n} \mathbf{J}_i = 0 \tag{4.14}$$

The effective binary diffusion coefficient $\mathscr{D}_{im}$ for compound i in a multi component system is defined by Fick's first law which can be written

$$\mathbf{J}_i = -c \, \mathscr{D}_{im} \nabla x_i \tag{4.15}$$

where c is the total concentration. If the binary diffusion coefficients $\mathscr{D}_{ij}$ in an ideal system are independent of the composition, the *Stefan-Maxwell equation* is valid which, in this case, can be written

$$-c \nabla x_i = \sum_{j=1}^{n} \frac{x_j \mathbf{J}_i - x_i \mathbf{J}_j}{\mathscr{D}_{ij}} \tag{4.16}$$

Then, $\mathscr{D}_{im}$ can be calculated from the following equation.

$$\frac{\mathbf{J}_i}{\mathscr{D}_{im}} = \sum_{j=1}^{n} \frac{x_j \mathbf{J}_i - x_i \mathbf{J}_j}{\mathscr{D}_{ij}} \tag{4.17}$$

Let $\mathbb{I} = \{1, 2, 3, \ldots, n\}$ and $\mathbb{J}_i = \{j \in \mathbb{I} \,|\, j \neq i\}$. Then, Eq. (4.17) can be written

$$\frac{\mathbf{J}_i}{\mathscr{D}_{im}} = \mathbf{J}_i \sum_{j \in \mathbb{J}_i} \frac{x_j}{\mathscr{D}_{ij}} - x_i \sum_{j \in \mathbb{J}_i} \frac{\mathbf{J}_j}{\mathscr{D}_{ij}} \tag{4.18}$$

Assume that $x_i \ll 1$. Then, Eq. (4.18) can approximately be written

$$\frac{1}{\mathscr{D}_{im}} \approx \sum_{j \in \mathbb{J}_i} \frac{x_j}{\mathscr{D}_{ij}} \tag{4.19}$$

Assume that all $\mathscr{D}_{ij}$ are equal. Then, Eq. (4.18) can be written

$$\mathscr{D}_{im} = \mathscr{D}_{ij} \tag{4.20}$$

The first term on the right side in Eq. (4.18) is a vector which has the same direction as the vector on the left side in the equation. This means that the second term on the right side in Eq. (4.18) also is a vector which has the same direction as the vector on the left side. Assume that $\mathscr{D}_{im} \approx \mathscr{D}_{ij}$. Then, Eq. (4.14) means that

$$\frac{\mathbf{J}_i}{\mathscr{D}_{im}} + \sum_{j\in\mathbb{J}_i} \frac{\mathbf{J}_j}{\mathscr{D}_{ij}} \approx 0 \tag{4.21}$$

Then, Eq. (4.18) can approximately be written

$$\frac{1-x_i}{\mathscr{D}_{im}} \approx \sum_{j\in\mathbb{J}_i} \frac{x_j}{\mathscr{D}_{ij}} \tag{4.22}$$

Equation (4.22) often gives a good estimation of $\mathscr{D}_{im}$. Otherwise, Eq. (4.18) must be included in a system of equations where $\mathscr{D}_{im}$ and $\mathbf{J}_i$ are system variables. The velocity $\mathbf{v}^*$ is defined by

$$c\,\mathbf{v}^* = \mathbf{N} \tag{4.23}$$

The velocity $\mathbf{v}$ is defined by

$$\rho\,\mathbf{v} = \mathbf{G} \tag{4.24}$$

where $\mathbf{G}$ is the mass flux in the fluid. $\mathbf{G}$ can be written

$$\mathbf{G} = \sum_{i=1}^{n} M_i\,\mathbf{N}_i \tag{4.25}$$

where M_i is the molar weight of compound i. If one solves the flow equation for the fluid in a multi component system, the solution gives the velocity $\mathbf{v}$. This means that the following equation is a part of the equation system.

$$\rho\,\mathbf{v} = \sum_{i=1}^{n} M_i\,\mathbf{N}_i \tag{4.26}$$

4.5 The Mass Transport Equation

Assume that the effective binary diffusivity $\mathscr{D}_{im}$ and the total concentration c are independent of the space coordinates. Then, the mass transport equation can be written

$$\mathcal{D}_{im} \nabla^2 x_i + \frac{q_i}{c} = \frac{\partial x_i}{\partial t} + \nabla \cdot \left(x_i \, \mathbf{v}^* \right) \tag{4.27}$$

Equation (4.27) is obtained from Eqs. (4.2), (4.12), (4.15) and (4.23). Equation (4.27) is very similar to Eq. (3.26). The difference is the second term on the right side. If $\mathbf{v}^* = \mathbf{v}$ and $\nabla \cdot \mathbf{v} = 0$, the equations are completely analogous. In this case, Eq. (4.27) can be written

$$\mathcal{D}_{im} \nabla^2 x_i + \frac{q_i}{c} = \frac{\partial x_i}{\partial t} + \nabla x_i \cdot \mathbf{v} \tag{4.28}$$

It is very common that $\mathbf{v}^* \approx \mathbf{v}$. For incompressible flow, $\nabla \cdot \mathbf{v} = 0$. This means that solutions to heat transfer problems approximately can be used for solution of mass transfer problems. If $q_i = 0$ and $\mathbf{v} = 0$, Eq. (4.28) can be written

$$\mathcal{D}_{im} \nabla^2 x_i = \frac{\partial x_i}{\partial t} \tag{4.29}$$

Equation (4.29) is called Fick's second law.

4.6 Mass Transfer Coefficient

Consider a fixed object, bounded by the closed surface S, which is located in a flowing fluid in an infinite space where there are no other objects. Assume that the concentration c_i of compound i is constant and equal to c_{ib} at infinite distance from the object and that the concentration of compound i at the surface S of the object is constant and equal to c_{is}. The molar flux $\mathbf{N}_i$ of compound i at the surface S can be expressed by Fick's first law (4.15). The mass transfer coefficient β_i for compound i is defined

$$\int_S \mathbf{N}_i \cdot \mathbf{S} = \beta_i \, A \, (c_{is} - c_{ib}) \tag{4.30}$$

where A is the area of the surface S and c_{ib} is the *bulk concentration* of compound i. The unit normal $\mathbf{n}$ is directed out from the surface S. Note that the molar flux $\mathbf{N}$ in the general case does not need to be directed in the same direction as $\mathbf{n}$.

Consider a straight channel with constant cross section which contains an incompressible flowing fluid with constant viscosity μ. Assume that the flow profile $\mathbf{v}$ is independent of the axial coordinate x and the time t. Further, assume that S_x is a plane cross section which is perpendicular to the x-axis and bounded by the channel wall. The *bulk concentration* c_{ib} of compound i is defined

$$\int_{S_x} c_i \, \mathbf{v} \cdot \mathbf{S} = c_{ib} \int_{S_x} \mathbf{v} \cdot d\mathbf{S} \tag{4.31}$$

Consider a segment with the length Δx along the x-axis. Assume that ΔS is the surface of the channel in the segment and that the surface concentration c_{is} is constant on the boundary for constant x. The mass transfer coefficient β_i between the channel surface and the fluid is defined

$$\lim_{\Delta A \to 0} \frac{1}{\Delta A} \int_{\Delta S} \mathbf{N}_i \cdot d\mathbf{S} = \beta_i \, (c_{is} - c_{ib}) \tag{4.32}$$

where ΔA is the area of the surface ΔS. The unit normal $\mathbf{n}$ is directed towards the center of the channel. In the both cases above, the bulk concentration c_{ib} is defined in quite different way. If the space contains several objects, one can not define a bulk concentration unless the objects are far apart. The molar flow from a single object is influenced by the presence of the other objects. If the objects are grouped in a cluster, one can define the bulk concentration as the concentration of compound i at infinite distance from the cluster. Then, it is possible to define a common mass transfer coefficient for the cluster if the surface concentration on the objects is the same. In this case, the cluster can be considered as a single object. Consider a large number of uniform objects which are regularly placed in the space in such way that the flow pattern around all objects becomes equal. This case is very similar to the mass transfer in a channel. One can use Eq. (4.31) to define the bulk concentration in this case if S_x is a cross-sectional surface perpendicular to the expectation value of the velocity vector. The mass transfer coefficient will depend somewhat on where the cross cross-sectional surface S_x is positioned along the x-axis. Then, one must define the mass transfer coefficient as an expectation value. In a packed bed of particles, the particles are randomly distributed in the space. One can even in this case define the bulk concentration approximately by Eq. (3.29). There are some limitations of using mass transfer coefficients.

Consider an incompressible fluid with constant viscosity μ which contains compound i with constant value of $\mathscr{D}_{im}$. Assume that the fluid flows around a solid object with a given geometry and that the flow is time independent. For uniform objects, the geometry of an object can be described by only one length quantity D. D can, for example, be the diameter of a sphere. Further, assume that the object is oriented in a given way relative to the flow direction and that the flow at infinite distance from the object is homogeneous with constant velocity v_0. Then, one can introduce dimensionless space coordinates (ξ_x, ξ_y, ξ_z), dimensionless velocities $\mathbf{v}'$ and dimensionless concentrations c_i' in the following way.

$$(\xi_x, \xi_y, \xi_z) = \frac{\mathbf{r}}{D} \qquad \mathbf{r} = (x, y, z) \tag{4.33}$$

$$\mathbf{v}' = \frac{\mathbf{v}}{v_0} \tag{4.34}$$

$$c_i' = \frac{c_i - c_{ib}}{c_{is} - c_{ib}} \tag{4.35}$$

$\mathbf{r}$ is the space coordinate vector. If $q_i = 0$, Eq. (4.28) can be written

$$\frac{\partial^2 c_i'}{\partial \xi_x^2} + \frac{\partial^2 c_i'}{\partial \xi_y^2} + \frac{\partial^2 c_i'}{\partial \xi_z^2} = Re\, Sc_i \left(v_1' \frac{\partial c_i'}{\partial \xi_x} + v_2' \frac{\partial c_i'}{\partial \xi_y} + v_3' \frac{\partial c_i'}{\partial \xi_z} \right) \qquad (4.36)$$

$$Re = \frac{\rho\, v_0 D}{\mu} \qquad (4.37)$$

$$Sc_i = \frac{\mu}{\rho\, \mathcal{D}_{im}} \qquad (4.38)$$

where Re is the *Reynolds number* and Sc_i is the *Schmidt number* for compound i. The value of c_i' is equal to 1 on the surface of the object and equal to 0 at infinite distance from the object. This means that the boundary condition for c_i' is independent of c_{is} and c_{ib}. Equation (2.72) means that the flow field v_i' in the variable space (ξ_x, ξ_y, ξ_z) only depends on the Reynolds number Re for an object with given shape. Then, Eq. (4.36) means that the concentration field c_i' in the variable space (ξ_x, ξ_y, ξ_z) only depends on Re and Sc_i for an object with given shape. Fick's first law (4.15) means that Eq. (4.30) can be written

$$\beta_i\, A\, (c_{is} - c_{ib}) = -\mathcal{D}_{im} \int_S \nabla c_i \cdot d\mathbf{S}$$

$$= -\mathcal{D}_{im}\, D\, (c_{is} - c_{ib}) \int_{S'} \nabla' c_i' \cdot d\mathbf{S}' \qquad (4.39)$$

where $\mathbf{S}'$ is the surface of the object in the variable space (ξ_x, ξ_y, ξ_z), $d\mathbf{S} = D^2 d\mathbf{S}'$ and $\nabla' = (\partial/\partial \xi_x, \partial/\partial \xi_y, \partial/\partial \xi_z)$. Equation (3.39) can also be written

$$\frac{\beta_i\, A}{\mathcal{D}_{im}\, D} = -\int_{S'} \nabla' c_i' \cdot d\mathbf{S}' \qquad (4.40)$$

The right side in Eq. (4.40) only depends on Re and Sc_i. The area A of the outer surface S of the object is proportional to D^2 for an object with given shape. This means that the quantity $\beta_i D/\mathcal{D}_{im}$ is a function of Re and Sc_i. The quantity $\beta_i D/\mathcal{D}_{im}$ is called the *Sherwood number* Sh_i. Then, Eq. (4.40) can be written

$$\frac{\beta_i\, D}{\mathcal{D}_{im}} = Sh_i(Re, Sc_i) \qquad (4.41)$$

4.7 Analytical Solutions to the Mass Transport Equation

This section contains analytical solutions to the mass transport equation in some simple geometries. These solutions are based on the analytical solutions to the Navier-Stokes equation which are described in Chap. 2. The mass transport equation (4.27) is nearly analogous to the heat transport Eq. (3.26). Equations (4.28) and (3.26) are completely analogous. Chapter 3 contains analytical solutions to the heat transport equation. These solutions can be described as a connection between the Nusselt number Nu, the Reynolds number Re and the Prandtl number Pr. These connections can also be used to determine how the Sherwood number Sh_i depends on the Reynolds number Re and the Schmidt number Sc_i for the same geometries as in Chap. 3. The section also contains some correlations for mass transfer when there are no analytical solutions.

4.7.1 Mass Transfer Around a Sphere

The Navier-Stokes equation can be solved analytically for creep flow around a sphere. This solution is described in Sect. 2.7.1. The solution is only valid for Reynolds numbers which are less than 0.1. At such low values of the Reynolds number, the flow around the sphere will have little effect on the mass transfer. Therefore, it is not particularly meaningful to try to find an analytical solution for the concentration field around a sphere in the flow pattern which is described in Sect. 2.7.1. It is only meaningful to solve the mass transport equation analytically for stagnant flow in this case. Assume that $q_i = 0$ in Eq. (4.28) and D is the diameter of the sphere. Then, Eq. (3.43) means that the Sherwood number $Sh_i = \beta_i D / \mathcal{D}_{im}$ can be written

$$Sh_i = 2 \tag{4.42}$$

See also Sect. 3.5.1. According to Eq. (4.41), Sh_i is a function of Re and Sc_i. The following correlation is taken from *Transport Phenomena by Beek & Muttzall* [1].

$$Sh_i = 2 + 1.3 \cdot Sc_i^{0.15} + 0.66 \cdot Re^{0.50} \cdot Sc_i^{0.33} \quad 1 < Re < 10^4 \tag{4.43}$$

The following correlation is taken from *Transport Phenomena by Bird, Stewart & Lightfoot* [2].

$$Sh_i = 2 + 0.60 \cdot Re^{1/2} \cdot Sc_i^{1/3} \quad 1 < Re < 5 \cdot 10^4 \tag{4.44}$$

In the both cases, Re and Sc_i are defined according to Eqs. (4.37) and (4.38) where D is the diameter of the sphere.

4.7.2 Mass Transfer in a Cylindrical Tube

Consider a cylindrical tube which contains a flowing incompressible fluid with constant viscosity μ where the fluid contains compound i with constant value of $\mathcal{D}_{im}$. Assume that the radial molar flux of compound i is constant on the inner surface of the cylinder. Then, Eq. (3.54) means that the Sherwood number $Sh_i = \beta_i D / \mathscr{D}_{im}$ for laminar flow can be written

$$Sh_i = \frac{48}{11} \approx 4.36 \tag{4.45}$$

For fully developed laminar flow in a tube there are no acceleration forces in the flow. This means that Sh_i is independent of Re and Sc_i for constant molar flow on the tube wall. Equation (4.45) is valid for constant molar flux on the tube wall. There are reports in the literature that $Sh_i = 48/11$ for constant molar flux on the tube wall and $Sh_i = 3.66$ for constant wall concentration. The latter value of Sh_i is approximately valid if the bulk concentration c_{ib} varies moderately with z. See also Sect. 3.5.2.

The following correlations are taken from *Transport Phenomena by Bird, Stewart & Lightfoot* [3].

$$Sh_i = 7 + 0.025 \, (Re \, Sc_i)^{0.8} \tag{4.46}$$
$$Sh_i = 5 + 0.025 \, (Re \, Sc_i)^{0.8} \tag{4.47}$$

The both correlations are valid for turbulent flow. Equation (4.46) is valid for constant molar flux on the tube wall and Eq. (4.47) is valid for constant concentration on the tube wall. Transition to turbulent flow occurs when $Re \approx 2100$.

4.7.3 Mass Transfer in a Rectangular Channel

Section 3.5.3 describes heat transfer in a rectangular channel with the width $2a$ and the height $2b$. The center line of the channel coincides with the z-axis in a Cartesian coordinate system as in Fig. 2.3. The solution assumes that the temperature is constant and independent of x and y at the boundary for constant z and that the heat flux on the boundary is independent of z for constant x and y. This condition can be satisfied if the heat transfer coefficient in the surrounding solid material is infinitely high in the x- and y-coordinate direction and zero in the z-coordinate direction. In this case, the condition is satisfied if it occurs a heat generation in the solid material around the channel which is independent of z. In the case of mass transfer in a rectangular channel, it is difficult or impossible to satisfy the condition that the concentration on the boundary is constant and independent of x and y for constant z while the molar flux is constant on the boundary and independent of z for constant x and y. If the molar flux on the surface is constant and independent of x, y and z, the concentration on the boundary will depend on x and y for constant z. If the concentration on the surface is constant and independent of x, y and z, the molar flux on the boundary will depend on z for constant x and y. It is only possible to define a mass transfer coefficient in a channel if the concentration is constant and independent of x and y at the boundary for constant z.

Table 4.1 The Sherwood number Sh_i in a rectangular channel as function of a/b

a/b	Sh_i
1.0	3.608
0.9	3.620
0.8	3.664
0.7	3.750
0.6	3.895
0.5	4.123
0.4	4.472
0.3	4.990
0.2	5.738
0.1	6.785
0.0	8.235

Consider a rectangular channel with the width $2a$ and the height $2b$ which contains a flowing incompressible fluid with constant viscosity μ where the fluid contains compound i with constant value of $\mathscr{D}_{im}$. Assume that the center line of the channel coincides with the z-axis in a Cartesian coordinate system as in Fig. 2.3. Assume that the flow is stationary and independent of the coordinate z. This is valid for a fully developed flow profile. Assume that the concentration on the boundary is constant and independent of x and y for constant z and that the molar flux on the boundary for constant x and y is constant and independent of z. Then, Eq. (3.77) means that the Sherwood number $Sh_i = \beta_i D_h / \mathscr{D}_{im}$ for laminar flow can be written

$$Sh_i = \frac{\beta_i D_h}{\mathscr{D}_{im}} = \left(\frac{8}{\pi}\right)^2 \frac{\Sigma^2}{\Sigma_T} \frac{a\,b}{(a+b)^2} \tag{4.48}$$

where Σ is obtained from Eq. (2.123) and Σ_T is obtained from Eq. (3.76). The hydraulic diameter D_h is obtained from Eq. (2.124). Equation (4.48) is valid for fully developed laminar flow. Table 4.1 shows Sh_i for different values of a/b. See also Sect. 3.5.3.

4.7.4   Mass Transfer in a Slit

Consider a slit with the width $2a$ which contains an incompressible fluid with constant viscosity μ where the fluid contains compound i with constant value of $\mathscr{D}_{im}$. Assume that the center surface of the slit coincides with the y-z-plane in a Cartesian coordinate system as in Fig. 2.4 and that the slit has infinite extension in the y-z-plane. Assume that the flow is stationary and independent of the coordinates y and z and that the flow is directed in the z-coordinate direction. Assume that the molar flux is constant on the inner surface of the slit. Then, Eq. (3.88) means that the Sherwood number $Sh_i = \beta_i D_h / \mathscr{D}_{im}$ for laminar flow can be written

$$Sh_i = \beta_i \frac{4a}{\mathscr{D}_{im}} = \frac{140}{17} \approx 8.235 \tag{4.49}$$

This value is consistent with the value of Sh_i in Table 4.1 when $a/b = 0$. According to Eq. (2.136), the hydraulic diameter D_h is equal to $4a$ for an infinite slit with the width $2a$. There are reports in the literature that $Sh_i = 140/17$ for constant molar flux on the wall and $Sh_i = 7.54$ for constant concentration on the wall. The latter value of Sh_i is approximately valid if the bulk concentration c_{ib} varies moderately with z. See also Sect. 3.5.4.

4.8 Empirical Correlations for Mass Transfer

4.8.1 Mass Transfer in a Packed Bed of Particles

The bulk concentration of compound i in a packed bed of particles which contains a flowing incompressible fluid can approximately be defined by Eq. (4.31). See Sect. 4.6. The Sherwood number Sh_i is a function of the Reynolds number Re and the Schmidt number Sh_i. See Eq. (4.41). This function can be determined by fitting of some empirical expression to experimental data. The following correlation is taken from *Transport Phenomena* by Bird, Stewart & Lightfoot [4].

$$Sh_i = j_D \, Re \, Sc_i^{1/3} \tag{4.50}$$

$$j_D = 0.91 \, Re^{-0.51} \, \Phi \quad (Re < 50) \tag{4.51}$$

$$j_D = 0.61 \, Re^{-0.41} \, \Phi \quad (Re > 50) \tag{4.52}$$

$$Re = \frac{G}{S_v \mu \, \Phi} \tag{4.53}$$

$$Sc_i = \frac{\mu}{\rho \, \mathscr{D}_{im}} \tag{4.54}$$

G is the superficial mass flow, S_v is the specific area of the bed, μ is the dynamic viscosity of the fluid, $\mathscr{D}_{im}$ is the effective binary diffusion coefficient for compound i, ρ is the density of the fluid and Φ is the shape factor of the particles. Table 4.2 shows the value of Φ for different particle shapes.

Table 4.2 The shape factor Φ for different particle shapes

Particle shape	Φ
Spheres	1.00
Cylinders	0.91
Flakes	0.86
Raschig rings	0.79
Partition rings	0.67
Berl-saddles	0.80

Colburn's j_D-factor is defined

$$j_D = \frac{\beta_i \, \rho}{G} \, Sc_i^{2/3}$$ (4.55)

where β_i is the mass transfer coefficient for compound i. The mass transfer coefficient β_i is obtained from the following equation.

$$\beta_i = Sh_i \, S_v \, \mathscr{D}_{im} \, \Phi$$ (4.56)

Creep flow in a packed bed means that the flow is so low that the acceleration forces in the fluid are negligible compared with the viscous forces. For creep flow ($Re \lesssim 0.1$) in a packed bed, Sh_i is independent of Re. This means that the correlation for for Sh_i above is not valid for very low values of Re. At creep flow $j_D \propto Re^{-1}$.

4.8.2 Mass Transfer in a Package of Wire Screens

The bulk concentration of compound i in a package of wire screens which contains a flowing incompressible fluid can approximately be defined by Eq. (4.31). See Sect. 4.6. The Sherwood number Sh_i is a function of the Reynolds number Re and the Schmidt number Sh_i. See Eq. (4.41). This function can be determined by fitting of some empirical expression to experimental data. The following correlation is taken from an article by *Satterfield & Cortez* [5].

$$Sh_i = j_D \, Re \, Sc_i^{1/3}$$ (4.57)

$$j_D = \frac{0.874}{\varepsilon_b} \left[4 \left(1 - \varepsilon_b \right) Re \right]^{-0.645}$$ (4.58)

$$Re = \frac{G}{S_v \, \mu}$$ (4.59)

$$Sc_i = \frac{\mu}{\rho \, \mathscr{D}_{im}}$$ (4.60)

ε_b is the porosity of the bed. The correlation above assumes that the wire screens are placed perpendicular to the flow. The mass transfer coefficient β_i is obtained from the following equation.

$$\beta_i = Sh_i \, S_v \, \mathscr{D}_{im}$$ (4.61)

Moreover, the same notations and equations are valid as in Sect. 4.8.1. See also Sect. 4.8.1.

4.9 Mass Transport in Porous Particles

In heterogeneous catalysis, the solid catalyst is distributed on the surface of a catalyst support, formed as a porous particle. Assume that the mean molecular path in the fluid is much less than the pore diameter. Then, the molar flux $\mathbf{N}_i$ of compound i in the particle can be written

$$\mathbf{N}_i = -c\,\mathcal{D}_{ie}\nabla x_i + x_i\mathbf{N} \tag{4.62}$$

$$\mathcal{D}_{ie} = \frac{\varepsilon_p}{f}\,\mathcal{D}_{im} \tag{4.63}$$

where x_i is the molar fraction of compound i, c is the total concentration, $\mathcal{D}_{ie}$ is the effective diffusion coefficient of compound i, ε_p is the porosity of the particle and f is the winding factor of the pores. $\mathbf{N}$ is obtained from Eq. (4.13). Assume that the particle is a sphere with the radius R where $\mathbf{N} = 0$, x_i is time independent and x_i only depends on the radial coordinate r in a spherical coordinate system. Then, Eq. (4.28) can be written

$$\mathcal{D}_{ie}\frac{1}{r}\frac{\partial^2}{\partial r^2}(r\,x_i) + \frac{q_i}{c} = 0 \tag{4.64}$$

Assume that there occurs a first order reaction with respect to compound i in such a way that q_i can be written

$$q_i = -k\,c\,x_i \tag{4.65}$$

where k is a rate constant. Then, Eq. (4.64) can be written

$$\frac{\partial^2}{\partial r^2}(r\,x_i) - \frac{k}{\mathcal{D}_{ie}}(r\,x_i) = 0 \tag{4.66}$$

The solution to Eq. (4.66) can be written

$$x_i = \frac{C_1}{r}e^{\phi_i r/R} + \frac{C_2}{r}e^{-\phi_i r/R} \tag{4.67}$$

$$\phi_i = R\sqrt{\frac{k}{\mathcal{D}_{ie}}} \tag{4.68}$$

where ϕ_i is the *Thiele modulus* for compound i. The condition that $x_i(r)$ must be limited in the origin means that $C_1 + C_2 = 0$. The boundary condition $x_i(R) = x_{is}$ means that $x_i(r)$ can be written

$$x_i(r) = x_{is}\frac{R}{r}\frac{\sinh(\phi_i r/R)}{\sinh(\phi_i)} \tag{4.69}$$

where x_{is} is the molar fraction of compound i on the outer surface of the sphere.

Notations

Roman characters

A Area (m^2).

a The half width of a slit or the half width in a rectangular channel (m).

b The half height in a rectangular channel (m).

c The total concentration (mol/m^3).

c_i The concentration of compound i (mol/m^3).

c_{ib} The bulk concentration of compound i (mol/m^3).

c_{is} The surface concentration of compound i (mol/m^3).

D Characteristic length (m).

D_h Hydraulic diameter (m).

D_{ij} Binary diffusion coefficient in a mixture of compound i and j (m^2/s). See Eq. (4.5).

$\mathscr{D}_{ie}$ Effective diffusion coefficient for compound i in a porous particle (m^2/s).

$\mathscr{D}_{ij}$ Binary diffusion coefficient in a mixture of compound i and j (m^2/s). See Eq. (4.3).

$\mathscr{D}_{im}$ Effective binary coefficient for compound i in a multi component system (m^2/s).

f The winding factor of the pores.

G Superficial mass flow (kg/s,m^2).

$\mathbf{G}$ Mass flux (kg/s,m^2).

$\mathbf{J}_i$ Molar flux of compound i caused by diffusion (mol/s,m^2).

j_D Colburn's j_D-factor.

k Rate constant (1/s).

M_i Molar weight of compound i (kg/mol).

$\mathbf{N}$ Total molar flux (mol/s,m^2).

$\mathbf{N}_i$ Molar flux of compound i caused by convection and diffusion (mol/s,m^2).

n Number of compounds in a multi component system.

$\mathbf{n}$ The unit normal.

q_i Formation rate of compound i (mol/s,m^3).

Sc_i Schmidt number for compound i.

Sh_i Sherwood number for compound i.

S_v Specific area (1/m).

t Time (s).

$\mathbf{v}$ The velocity of the fluid based on mass flux (m/s).

$\mathbf{v}^*$ The velocity of the fluid based on molarflux (m/s).

x Space coordinate (m).

x_i Molar fraction of compound i.

y Space coordinate (m).

z Space coordinate (m).

Greek characters

β_i Mass transfer coefficient for compound i (m/s).
ε_b Porosity in a package of wire screens.
ε_p The porosity of the particle.
μ The dynamic viscosity of the fluid (kg/m,s).
ξ_x Dimensionless space coordinate (x/D).
ξ_y Dimensionless space coordinate (y/D).
ξ_z Dimensionless space coordinate (z/D).
ρ The density of the fluid (kg/m^3).
Φ The shape factor of the particles.
ϕ_i Thiele modulus for compound i.

References

1. W.J. Beek, K.M.K. Muttzall, *Transport Phenomena* (John Wiley & Sons Ltd, New York, 1975), p. 196, equation (III.68)
2. R.B. Bird, W.E. Stewart, E.N. Lightfoot, *Transport Phenomena* (John Wiley & Sons, New York, 1960), p. 409, equation (13.3-1)
3. R.B. Bird, W.E. Stewart, E.N. Lightfoot, *Transport Phenomena* (John Wiley & Sons, New York, 1960), p. 404, equation (13.2-21) and (13.2-22)
4. R.B. Bird, W.E. Stewart, E.N. Lightfoot, *Transport Phenomena* (John Wiley & Sons, New York, 1960), p. 411–412, equation (13.4-1) to (13.4-5)
5. C.N. Satterfield, D.H. Cortez, Mass transfer characteristics of woven-wire screen catalysts. Ind. Eng. Chem. Fundam. **9**(4), 613 (1970)

Erratum to: Elementary Mathematics

Erratum to:
Chapter 1, in: P. Olsson, *Transport Phenomena*
in Newtonian Fluids - A Concise Primer,
SpringerBriefs in Continuum Mechanics,
DOI 10.1007/978-3-319-01309-1_1

There is a change in Eq. (1.8). The term "0" is deleted and equation should read as below:

$$\nabla \times \mathbf{F} = \begin{vmatrix} \widehat{\mathbf{x}}_1 & \widehat{\mathbf{x}}_2 & \widehat{\mathbf{x}}_3 \\ \frac{\partial}{\partial x_1} & \frac{\partial}{\partial x_2} & \frac{\partial}{\partial x_3} \\ F_1 & F_2 & F_3 \end{vmatrix}$$
$$= \widehat{\mathbf{x}}_1 \left(\frac{\partial F_3}{\partial x_2} - \frac{\partial F_2}{\partial x_3} \right) + \widehat{\mathbf{x}}_2 \left(\frac{\partial F_1}{\partial x_3} - \frac{\partial F_3}{\partial x_1} \right) + \widehat{\mathbf{x}}_3 \left(\frac{\partial F_2}{\partial x_1} - \frac{\partial F_1}{\partial x_2} \right)$$

$$(1.8)$$

The online version of the original chapter can be found under DOI 10.1007/978-3-319-01309-1_1

P. Olsson (✉)
Arvid Lindmansgatan 2A, 417 26 Göteborg, Sweden
e-mail: po.systemanalys@telia.com

P. Olsson, *Transport Phenomena in Newtonian Fluids - A Concise Primer,*
SpringerBriefs in Continuum Mechanics,
DOI: 10.1007/978-3-319-01309-1_5, © The Author(s) 2014

Erratum to: Transport Phenomena in Newtonian Fluids - A Concise Primer

Erratum to:
P. Olsson, *Transport Phenomena in Newtonian Fluids - A Concise Primer*, SpringerBriefs in Continuum Mechanics, DOI 10.1007/978-3-319-01309-1

There is a change in Eq. (1.8). The term "0" is deleted and equation should read as below:

$$\nabla \times \mathbf{F} = \begin{vmatrix} \widehat{\mathbf{x}}_1 & \widehat{\mathbf{x}}_2 & \widehat{\mathbf{x}}_3 \\ \frac{\partial}{\partial x_1} & \frac{\partial}{\partial x_2} & \frac{\partial}{\partial x_3} \\ F_1 & F_2 & F_3 \end{vmatrix}$$

$$= \widehat{\mathbf{x}}_1 \left(\frac{\partial F_3}{\partial x_2} - \frac{\partial F_2}{\partial x_3} \right) + \widehat{\mathbf{x}}_2 \left(\frac{\partial F_1}{\partial x_3} - \frac{\partial F_3}{\partial x_1} \right) + \widehat{\mathbf{x}}_3 \left(\frac{\partial F_2}{\partial x_1} - \frac{\partial F_1}{\partial x_2} \right)$$

(1.8)

There are few additional amendments apart from Eq. 1.8 in this Book, as given below:

- Equations 2.14, 4.30, 4.31 should read as:

$$-\int_S \rho\, v_i\, dS = \frac{\partial}{\partial t} \int_V \rho\, dV$$

(2.14)

The online version of the original book can be found under DOI 10.1007/978-3-319-01309-1

P. Olsson (✉)
Arvid Lindmansgatan 2A, 417 26 Göteborg, Sweden
e-mail: po.systemanalys@telia.com

P. Olsson, *Transport Phenomena in Newtonian Fluids - A Concise Primer*,
SpringerBriefs in Continuum Mechanics,
DOI: 10.1007/978-3-319-01309-1_6, © The Author(s) 2014

- $$\int_S \mathbf{N}_i \cdot d\mathbf{S} = \beta_i A \left(c_{is} - c_{ib} \right) \tag{4.30}$$

- $$\int_{S_x} c_i \, \mathbf{v} \cdot d\mathbf{S} = c_{ib} \int_{S_x} \mathbf{v} \cdot d\mathbf{S} \tag{4.31}$$

Erratum to: Transport Phenomena in Newtonian Fluids - A Concise Primer

Per Olsson

Erratum to:
P. Olsson, *Transport Phenomena in Newtonian Fluids - A Concise Primer*, SpringerBriefs in Continuum Mechanics, https://doi.org/10.1007/978-3-319-01309-1

The original version of this book was revised. The following corrections have been carried out in Chaps. 2, 3, and 4:

Chapter 2:

p. 41: There is an error in the paragraph after equation (2.194), second sentence. The correct sentence is:

If the solution $\Psi(x, y)$ can be written as Eq. (2.189), Eq. (2.194) must be valid.

p. 42: A plus sign is missing in equation (2.203).

The correct equation (2.203) is:

$$\mu \left(\bar{v}_{i,jj} + v'_{i,jj} \right)$$
$$= \rho \left(\frac{\partial \bar{v}_i}{\partial t} + \bar{v}_j \bar{v}_{i,j} - g_i \right) + \bar{p}_{,i}$$
$$+ \rho \left(\frac{\partial v'_i}{\partial t} + v'_j v'_{i,j} + \bar{v}_j v'_{i,j} + v'_j \bar{v}_{i,j} \right) + p'_{,i}$$

The updated online version of Chap. 2 can be found at
https://doi.org/10.1007/978-3-319-01309-1_2

The updated online version for this book can be found at
https://doi.org/10.1007/978-3-319-01309-1

Chapter 3:

p. 60: There is an error in the paragraph after equation (3.17). The correct paragraph is: where $\widehat{H}_0$ is the enthalpy at the temperature T_0 and c_p is the heat capacity. If there occur chemical reactions in the fluid, $\widehat{H}_0$ will change with time in a point which follows the flow. This gives rise to a heat generation q which depends on the material time derivative of $\widehat{H}_0$ in the following way.

The updated online version of Chap. 3 can be found at
https://doi.org/10.1007/978-3-319-01309-1_3

Chapter 4:

p. 84: There is an error in equation (4.43). The correct equation (4.43) is:

$$Sh_i = 2 + 1.3 \cdot Sc_i^{0.15} + 0.66 \cdot Re^{0.50} \cdot Sc_i^{0.33} \qquad 1 < Re < 10^4$$

The updated online version of Chap. 4 can be found at
https://doi.org/10.1007/978-3-319-01309-1_4

Index

Printed in the United States
By Bookmasters